Cooperative Connected and Automated Mobility (CCAM)

Cooperative Connected and Automated Mobility (CCAM): Technologies and Applications

Special Issue Editor

Joaquim Ferreira

MDPI • Basel • Beijing • Wuhan • Barcelona • Belgrade

Special Issue Editor
Joaquim Ferreira
Campus Universitário
de Santiago
Portugal

Editorial Office
MDPI
St. Alban-Anlage 66
4052 Basel, Switzerland

This is a reprint of articles from the Special Issue published online in the open access journal *Electronics* (ISSN 2079-9292) in 2019 (available at: https://www.mdpi.com/journal/electronics/special_issues/Cooperative_Connected).

For citation purposes, cite each article independently as indicated on the article page online and as indicated below:

LastName, A.A.; LastName, B.B.; LastName, C.C. Article Title. *Journal Name* **Year**, *Article Number*, Page Range.

ISBN 978-3-03928-158-9 (Pbk)
ISBN 978-3-03928-159-6 (PDF)

Cover image courtesy of Joaquim Ferreira.

© 2020 by the authors. Articles in this book are Open Access and distributed under the Creative Commons Attribution (CC BY) license, which allows users to download, copy and build upon published articles, as long as the author and publisher are properly credited, which ensures maximum dissemination and a wider impact of our publications.
The book as a whole is distributed by MDPI under the terms and conditions of the Creative Commons license CC BY-NC-ND.

Contents

About the Special Issue Editor vii

Preface to "Cooperative Connected and Automated Mobility (CCAM): Technologies and
Applications" ix

Joaquim Ferreira
Cooperative, Connected and Automated Mobility (CCAM): Technologies and Applications
Reprinted from: *Electronics* **2019**, *8*, 1549, doi:10.3390/electronics8121549 1

Bao Liu, Feng Gao, Yingdong He and Caimei Wang
Robust Control of Heterogeneous Vehicular Platoon with Non-Ideal Communication
Reprinted from: *Electronics* **2019**, *8*, 207, doi:10.3390/electronics8020207 5

Xiao Zheng, Yuanfang Chen, Muhammad Alam and Jun Guo
Multi-Task Scheduling Based on Classification in Mobile Edge Computing
Reprinted from: *Electronics* **2019**, *8*, 938, doi:10.3390/electronics8090938 21

David Franco, Marina Aguado and Nerea Toledo
An Adaptable Train-to-Ground Communication Architecture Based on the 5G Technological
Enabler SDN
Reprinted from: *Electronics* **2019**, *8*, 660, doi:10.3390/electronics8060660 34

Marilisa Botte, Luigi Pariota, Luca D'Acierno and Gennaro Nicola Bifulco
An Overview of Cooperative Driving in the European Union: Policies and Practices
Reprinted from: *Electronics* **2019**, *8*, 616, doi:10.3390/electronics8060616 46

Xiaofeng Liu, Arunita Jaekel
Congestion Control in V2V Safety Communication: Problem, Analysis, Approaches
Reprinted from: *Electronics* **2019**, *8*, 540, doi:10.3390/electronics8050540 71

João Almeida, João Rufino, Muhammad Alam and Joaquim Ferreira
A Survey on Fault Tolerance Techniques for Wireless Vehicular Networks
Reprinted from: *Electronics* **2019**, *8*, 1358, doi:10.3390/electronics8111358 95

About the Special Issue Editor

Joaquim Ferreira holds a Ph.D in Informatics Engineering from the University of Aveiro (2005). Currently, he is adjunct professor at the University of Aveiro and researcher at the Telecommunications Institute (IT). His research interests include connected and automated vehicles (CAV), dependable distributed systems, fault-tolerant real-time communications, wireless vehicular communications and medium access control protocols. He was principal investigator, local coordinator or participant in over 15 funded national and international research projects. He is a senior member of IEEE, has participated in more than 30 conferences of scientific committees and has served as guest editor of several journals. He currently coordinates two connected mobility projects: Celtic Next SARWS and P2020 PASMO. He is also coordinating the participation of IT in two other connected mobility projects: P2020 TRUST and H2020 5G-MOBIX.

Preface to "Cooperative Connected and Automated Mobility (CCAM): Technologies and Applications"

Cooperative connected and automated mobility (CCAM) has the potential to reshape the transportation ecosystem in a revolutionary way. Transportation systems will be safer, more efficient and more comfortable. Cars are going to be the third living space, as passengers will have the freedom to use their car to live, work and travel. Despite the massive effort devoted, both by academia and industry, to developing connected and automated vehicles, there are still many issues to be addressed, including not only scientific and technological, but also regulatory and political issues.

This book, mostly centered on the scientific and technological aspects of CCAMs, features seven articles highlighting recent advances of the state of the art in different CCAM technologies. The field of cooperative connected and automated mobility is rather vast and multidisciplinary, and as a consequence many key aspects of CCAM technology were not addressed.

I would like to thank all authors who have contributed their work to this book.

Joaquim Ferreira
Special Issue Editor

Editorial

Cooperative, Connected and Automated Mobility (CCAM): Technologies and Applications

Joaquim Ferreira

Instituto de Telecomunicações/ESTGA, Universidade de Aveiro, Campus Universitário de Santiago, 3810-193 Aveiro, Portugal; jjcf@ua.pt

Received: 28 November 2019; Accepted: 10 December 2019; Published: 16 December 2019

1. Introduction

The advent of cooperative connected and automated mobility (CCAM) has the potential to fundamentally change the mobility paradigm towards mobility as a service, contributing to more safe, efficient and comfortable transportation systems. To materialize this visionary scenario, with a great societal impact, different aspects are involved, ranging from regulatory issues, standardization, certification and technological issues.

Current vehicles can already be considered to be connected devices, as many recent models are connected to external services, for infotainment, security, comfort and maintenance purposes. Furthermore, direct interaction among vehicles, pedestrians and the road infrastructure, using various radio technologies, is expected, in the short term, using consolidated and field-proven technologies. In parallel, the massive effort being made by many stakeholders towards automated driving, is mostly centered in the vehicles' sensors, as no automated actions are currently made based on data received from other vehicles or from the road infrastructure. However, it is commonly accepted that inter-vehicle cooperation will significantly enhance the overall traffic performance, safety and comfort. Vehicular communication will also foster the integration of automated vehicles within the intelligent transportation ecosystems, with automated vehicles communicating and cooperating with legacy vehicles, trams, bicycles, etc.

Vehicular communications, automated driving and cooperating transportation systems are increasingly being considered not only as complementary technologies, but also as the foundations of forthcoming visionary CCAM applications. To this end, regulatory entities, automotive OEMs, road operators, telecom operators and other stakeholders are converging to deploy large-scale infrastructures and field trials of cooperative intelligent transport systems to pave the way to cooperative, connected and automated vehicles.

2. The Present Issue

This Special Issue features seven articles highlighting recent advances of the state of the art in different technologies for cooperative, connected and automated mobility. The contents of these papers are introduced and summarised next.

Two papers address vehicular platooning, a key application of automated driving. The paper by Aslam et al. [1] presents empirical models relating the number of radio frequency (RF) neighbors that a platoon member, in a highway platoon-only scenario, observes, with relevant network performance metrics. The platoon performance largely depends on the quality of the communication channel, which in turn is highly influenced by the adopted medium access control protocol (MAC). Currently, VANETs use the IEEE 802.11p MAC, which follows a carrier sense multiple access with collision avoidance (CSMA/CA) policy, that is prone to collisions and degrades significantly with network load. This paper considers CSMA/CA, native IEEE 802.11p, and two TDMA-based overlay protocols, i.e., deployed over CSMA/CA, namely PLEXE-slotted and RA-TDMAp, to carry out extensive simulations with varying platoon sizes, number of occupied lanes and transmit power to deduce empirical models that provide estimates of average number of collisions per second and average busy time ratio. Simulation results show that these estimates can be obtained from observing the number of RF neighbors, i.e., number of distinct sources of the packets received by each vehicle per time unit. These estimates can enhance the online adaptation of distributed applications, particularly platooning control, to varying conditions of the communication channel.

The other paper on platooning, from Liu et al. [2] presents a state predictor based control strategy for heterogeneous vehicular platoon connected by non-ideal wireless communication. From the theoretical analysis and simulation results, it was concluded that both information delay and topological uncertainty caused by non-ideal wireless communication are critical to the stability and tracking performances of platoon, which need to be dealt with when synthesizing a platoon control system. Furthermore, when considering the information delay, besides the minimum topological eigenvalue the maximum one also affects the closed loop performance of platoon. Comparatively, the influence of minimum one can be ignored if only stability is taken into account. The proposed state predictor based control strategy can compensate for the information delay and the numerical approach based on LMI can find the required state feedback controller ensuring robust performance of platoon.

The paper from Zheng et al. [3] proposes a dynamic multi-task scheduling prototype to improve the limited resource utilization in the vehicular networks (VNET) assisted by mobile edge computing (MEC). To avoid conflicts between tasks when vehicles move, multi-task scheduling was regarded as a multi-objective optimization problem, with the goal to find the Pareto optimal solution. For this purpose, a Frank–Wolfe-based MGDA optimization algorithm was proposed and extended to the high-dimensional space. This paper concluded that Pareto optimal solution can be computed by an upper bound optimization.

The work presented in the paper from Franco et al. [4], addresses critical train communications, an increasingly important aspect of CCAM both for the automated trains and for the cooperation between trams and other vehicles. The paper presents an architecture based in 5G, SDN and on MPTCP to provide path diversity and end-to-end redundancy in order to contribute to a technology-independent and resilient communication service. In this architecture, SDN is a key enabler for addressing network flexibility and adaptability, due to its centralized control and its ability to deal with failures at runtime. Results indicate that the combination of MPTCP and SDN improves the train to ground communication performance indicators, compared to a legacy approach. MPTCP offers end-to-end redundancy by the aggregation of multiple access technologies, and SDN introduces path diversity to offer a resilient and reliable communication. Simulation results indicate that, when compared to a legacy communication architecture, the approach presented in this paper, demonstrate a clear improvement in the failover response time, while maintaining and even improving the uplink and downlink overall data rates.

The paper from Liu and Jaekel [5] presents a comprehensive survey of congestion control approaches for VANETs. Several relevant parameters and performance metrics, that can be used to evaluate these approaches, were identified and each approach was analyzed based a number of factors such as the type of traffic, whether it is proactive or reactive, and the mechanism for controlling congestion. It was concluded that there are still many challenges and open research problems for congestion control in V2V safety communications, namely joint power/rate control, improved awareness control, relative fairness and standardization.

Two review papers provide a systematic review of fault tolerance techniques for wireless vehicular networks [6] and an overview of the status of the policies and practices of cooperative driving in the European Union [7]. Fault tolerance techniques for wireless vehicular networks [6] are of utmost importance for safety-critical CCAM applications, such as cooperative self-driving cars and automated mobility in general. The paper by Almeida et al. [6] presents a systematic literature review, on this issue, of publications in journals and conferences proceedings, available through a set of different search databases. For that purpose, The PRISMA systematic method was adopted in order to identify the relevant papers for this survey. A comparison of the core features among the different solutions is presented, together with a brief discussion regarding the main drawbacks of the existing solutions, as well as the necessary steps to provide an integrated fault-tolerant approach to the future vehicular communications systems. It was observed an increasing trend in the recent years, with more protocols, mechanisms and architectures being proposed in order to enhance the dependability attributes of wireless vehicular networks. Nevertheless, there is still a shortage of strategies to completely fulfill the operation of dependable vehicular networks.

The paper by Botte et al. [7], mostly centered in roadside deployment activities, analyzes the policies and practices of cooperative driving in the European Union, aiming to shed a light on the current state of testing and deployment activities in the field at the start of 2019. This study is particularly timely given that the year 2019 was identified as the starting date for the deployment of mature services, and because the European Union legislation is paying great attention to the matter. An important conclusion of this paper is that the proliferation of autonomous vehicles, will significantly increase both the traveled distances and the number of trips. This will negatively impact the potential benefits of autonomous driving in terms of sustainability and congestion reduction.

3. Conclusions and Prospective Future Research Directions

This Special Issue has partially addressed some relevant CCAM-related technologies. There are, however, many more fundamental technologies for CCAM that were not considered here as, for example, cooperative perception, 5G and ultra-reliable low-latency communications, safety and security, advanced sensors, distributed control algorithms, machine learning, etc.

Over the last few years, both automotive OEMs and academia have dedicated a massive effort developing technologies for CCAM. Industry focus on CCAM has been mostly centered on automated driving, with some effort devoted to connectivity and very limited action towards cooperation between vehicles. Although there are still many open issues in the vehicle automation field, the automated cooperation between vehicles, e.g., for cooperative maneuvering, has the potential to dramatically increase the overall traffic efficiency and safety. Vehicle cooperation requires ultra-reliable low-latency communications, both short-range for, e.g., maneuver negotiation and long-range for, e.g., traffic management. Communication technologies supporting these requirements are already available, but the main impairments for automated inter-vehicle cooperation are trust and liability. In this context, there is still a wide range of prospective future research directions, in the various domains of CCAM: automated driving, vehicular networks and inter-vehicle cooperation.

Funding: This research received no external funding.

Acknowledgments: I would like to thank all of the authors who submitted articles to this Special Issue for their valuable contributions, and to all the reviewers who helped in the evaluation of the manuscripts and contributed to improve the quality of the papers. I am also grateful to the Electronics Editorial office staff who worked thoroughly to maintain the rigorous peer-review schedule and timely publication, and to the editorial board of MDPI Electronics, for their invitation to guest edit this Special Issue.

Conflicts of Interest: The author declares no conflict of interest.

References

1. Aslam, A.; Santos, P.S.F.A.L. Empirical Performance Models of MAC Protocols for Cooperative Platooning Applications. *Electronics* **2019**, *8*, 1334, doi:10.3390/electronics8111334. [CrossRef]
2. Liu, B.; Gao, F.H.Y.W.C. Robust Control of Heterogeneous Vehicular Platoon with Non-Ideal Communication. *Electronics* **2019**, *8*, 207, doi:10.3390/electronics8020207. [CrossRef]
3. Zheng, X.; Chen, Y.A.M.G.J. Multi-Task Scheduling Based on Classification in Mobile Edge Computing. *Electronics* **2019**, *8*, 938, doi:10.3390/electronics8090938. [CrossRef]
4. Franco, D.; Aguado, M.T.N. An Adaptable Train-to-Ground Communication Architecture Based on the 5G Technological Enabler SDN. *Electronics* **2019**, *8*, 660, doi:10.3390/electronics8060660. [CrossRef]
5. Liu, X.; Jaekel, A. Congestion Control in V2V Safety Communication: Problem, Analysis, Approaches. *Electronics* **2019**, *8*, 540, doi:10.3390/electronics8050540. [CrossRef]

6. Almeida, J.; Rufino, J.A.M.F.J. A Survey on Fault Tolerance Techniques for Wireless Vehicular Networks. *Electronics* **2019**, *8*, 1358, doi:10.3390/electronics8111358. [CrossRef]
7. Botte, M.; Pariota, L.D.L.B.G. An Overview of Cooperative Driving in the European Union: Policies and Practices. *Electronics* **2019**, *8*, 616, doi:10.3390/electronics8060616. [CrossRef]

© 2019 by the author. Licensee MDPI, Basel, Switzerland. This article is an open access article distributed under the terms and conditions of the Creative Commons Attribution (CC BY) license (http://creativecommons.org/licenses/by/4.0/).

Article

Robust Control of Heterogeneous Vehicular Platoon with Non-Ideal Communication

Bao Liu [1], Feng Gao [1,*], Yingdong He [2] and Caimei Wang [1]

1. School of Automotive Engineering, Chongqing University, Chongqing 400044, China; 20161102021@cqu.edu.cn (B.L.); 20183202001T@cqu.edu.cn (C.W.)
2. Mechanical Engineering, University of Michigan, Ann Arbor, MI 48109, USA; heyingd@umich.edu
* Correspondence: gaofeng1@cqu.edu.cn; Tel.: +86-189-9618-8196

Received: 5 November 2018; Accepted: 29 January 2019; Published: 12 February 2019

Abstract: The application of wireless communication to platooning brings such challenges as information delay and varieties of interaction topologies. To compensate for the information delay, a state predictor based control strategy is proposed, which transmits the future information of nodes instead of current values. Based on the closed loop dynamics of platoon with state predictor and feedback controller, a decoupling strategy is presented to analysis and design the platoon control system with lower order by adopting the eigenvalue decomposition of topological matrix. A numerical method based on LMI (Linear Matrix Inequality) is provided to find the required robust performance controller. Moreover, the influence of information delay on performance is studied theoretically and it is found that the tolerable maximum delay is determined by the maximum topological eigenvalue. The effectiveness of the proposed strategy is validated by several comparative simulations under various conditions with other methods.

Keywords: cooperative control; vehicular platoon; multi-agent system; communication delay; system decoupling

1. Introduction

The problem of traffic congestion, safety and pollution is getting increasingly serious [1]. Worldwide researchers, companies and governments are devoted to solving these issues and one of these technologies is called platooning, that is, a coordinated motion of a group of vehicles with the same destination and speed (referring to the leader) [2,3]. By this way, the air drag is reduced greatly, meanwhile the traffic flow is increased due to the smaller car-following distance. Many demo projects have already been carried out, for example, the Konvoi project in Germany [4], the founding of the California Partners for Advanced Transit and Highways program in USA [5], the Energy Intelligent Transport Systems project in Japan [6]. With these efforts, many challenges on practical application have been studied and overcome, such as vehicle dynamic control [7,8], vehicle-to-vehicle (V2V) communication [9,10], platoon controller design [11,12] and spacing policies [13].

Over the past 30 years, wireless communication has experienced rapid development, meanwhile lots of new V2V technologies appear, for example, vehicular ad-hoc network and dedicated short range communication [14,15]. Besides such assistance systems as collision warning at intersection, V2V has also been applied to platooning, which brings the benefit of better performances but leads to varieties of information topologies, for example, the bidirectional type (BD), two-predecessors following type (TPF), predecessor-leader following type (PLF) and so forth. [11,12]. Moreover compared with earlier radar-based system, V2V inevitably introduces other issues like information delay, packet dropout and so forth. From the previous studies, it is known that delay will cause platoon to be degraded and even string instable [16].

To achieve better performances, many advanced control approaches have been applied, such as linear state feedback control [17], optimal control [18], sliding mode control [19] and model predictive control, which require the information topology to be known in advance and unchanged [20]. When V2V is adopted, the topology becomes uncertain because of the environmental degradation of communication, which combined with information delay poses great challenges on platoon control system design. One way is to consider the behavior of platoon as a multi-agent system and the varieties of topologies is uniformly described by the graph theory. In particular, it is known that the consensus control performance is influenced by the Laplacian matrix [21,22]. Based on this approach, Zheng et al. studied the stability of homogeneous platoon based on the Routh–Hurwitz theory [23]. Moreover to deal with the uniform delay, Peters et al. provided a general linear control method [24] and Gong et al. established a newly sampled-data control method [25]. Both of them guarantee the string stability of homogeneous platoon with PLF. Gao et al. further presented an H_∞ control method for heterogeneous platoon to ensure robust stability, string stability and tracking ability simultaneously [26]. Considering heterogeneous delay, Bernardo et al. designed a distributed controller by treating it as a consensus issue [27]. This strategy maintains a stable platoon but the referenced velocity is constant and only PLF is applicable. Besides these advanced methods dealing with communication delay, some new wireless communication technologies, such as future 5G and LTE-V2X, have been studied and demonstrated to enhance communication itself [28,29]. These new technologies reduce the delay to a negligible level but today IEEE 802.11p and its related standard are already available [30].

The aforementioned work mostly focuses on one of the difficulties caused by V2V, that is, the topological uncertainties and heterogeneous communication delay. Consequently, the primary objective and main contribution of this paper are: 1) To deal with heterogeneous communication delay, a state predictor is designed to transmit future information of next sampling so that the negative effect of delay is compensated for. 2) Considering the topological uncertainty, eigenvalue decomposition of topological matrix and linear transformation are applied to decouple the platoon into multiple systems with lower order. 3) Based on the Lyapunov stability theory, a numerical way is presented to numerically solve a robust state feedback controller by LMI (Linear Matrix Inequality).

The reminder of paper is organized as follows: Section 2 introduces the studied problems and the closed-loop dynamics of the vehicular platoon is established in Section 3. Section 4 presents the decoupling design strategy and the theoretical analysis is conducted in Section 5. Comparative simulations are demonstrated in Sections 6 and 7 concludes the paper.

2. Problem Description

2.1. Notations and Definitions

Let $\mathbb{R}^{m \times n}$ be the set composed of real matrix with dimension $m \times n$, $\mathbf{0}$ denote the matrix with all entries being zero, $I_N \in \mathbb{R}^{N \times N}$ be the identity matrix and $1_N \in \mathbb{R}^N$ be the vector with all entries being 1. Notion $\|\cdot\|_2$ is the L_2 norm of signal and its induced norm is $\|\cdot\|_\infty$, the superscript "T" represents the transposition of matrix, "*" denotes the symmetric part in one matrix, "⊗" represent the Kronecker product, $\sigma(X)$ is the singular value of X, among which $\overline{\sigma}(X)$ and $\underline{\sigma}(X)$ are the maximum and minimum one respectively.

2.2. Problem Analysis Caused by V2V

As shown in Figure 1a, the studied platoon consists of $N + 1$ heterogeneous vehicles, that is, N followers (Indexed by $1, \cdots, N$) and one leader (Indexed by 0) sharing their state by V2V.

Because of the degradation of V2V, two issues arise naturally: (1) Uncertainties of topology caused by change of communication range and so forth. (See Figure 1a); (2) Information delay influenced by environments and distances. Figure 1b gives an example of the delay between two nodes, where node j is assumed to receive the state $x_i(kh)$ of node i, h is the sampling period and the delay is $t_d(k)$. At the $(k+1)$-th sample, node j calculates its control with the state $x_i(kh)$ at k-th sampling of node i, which introduces a time delay in the closed loop dynamics and is bad to control performances. As indicated by red line in Figure 1b, if all nodes transmit their predicted state at the future, such information delay can be reduced and even eliminated completely when the sampling time is known. Motivated by this idea and considering the fact that such global clock as GPS (Global Position System) can be used to synchronize the control period of platoon [16], a state predictor based platoon control system is proposed as shown in Figure 2, where K is the state feedback to be designed in Section 4.2. With this control strategy, all nodes transmit their predicted state at next sampling and control themselves by a predefined control cycle to compensate for the information delay. The topology uncertainty will be dealt with in Section 4 based on the closed loop dynamics of this predictor based platoon control system.

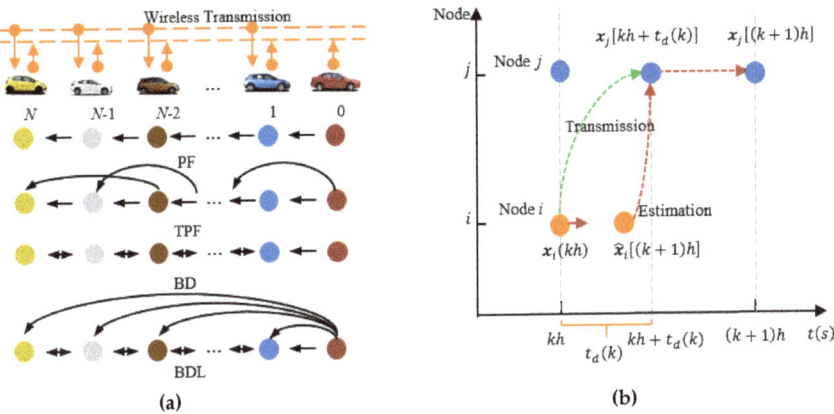

Figure 1. Problems arising from V2V. (a) Varieties of information flow topologies. (b) Fundamental of state predictor.

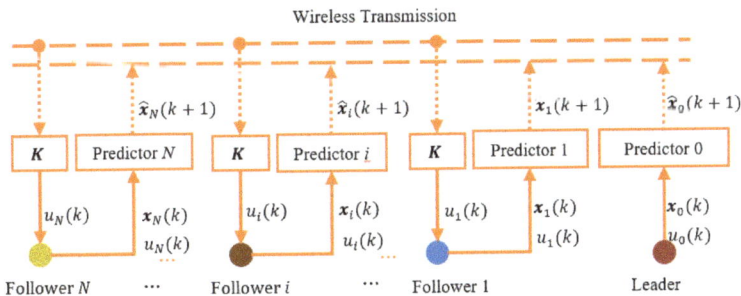

Figure 2. State predictor based platoon control system.

3. Closed-Loop Dynamical Model of Platoon

3.1. Vehicle Dynamical Model

Though the original vehicle dynamics is nonlinear, referring to [11,23,26] the following linear format is used to describe the dynamics of vehicle node, whose nonlinearities and environmental disturbances are compensated for by a lower dynamical controller [8]:

$$x_i(k+1) = A_i x_i(k) + B_i u_i(k), \quad i = 0, \ldots, N, \tag{1}$$

where A_i and B_i are the state and control matrices respectively, k denotes the sampling time, $u_i(k)$ is the control input, $x_i(k) = \begin{bmatrix} p_i(k) & v_i(k) & a_i(k) \end{bmatrix}^T$ where $p_i(k)$, $v_i(k)$ and $a_i(k)$ are the position, velocity and acceleration of node i. Note that the studied platoon is heterogeneous, that is, $A_0 \neq \cdots \neq A_N$ and $B_0 \neq \cdots \neq B_N$ but each vehicle known its own parameters. According to the control strategy shown by Figure 2 and (1), the state predictor is designed to be

$$\hat{x}_i(k+1) = A_i x_i(k) + B_i u_i(k), \; i = 0, \ldots, N, \tag{2}$$

where $\hat{x}_i(k)$ is the predicted state. Note that being different from followers controlled by the designed state-feedback controller, leader receives the command from others, such as driver or traffic management system whose dynamical characteristics are unknown.

3.2. Topology of Information Flow

To generate a unified description of varieties of topologies, the graph theory is used to model as a directed graph $G = (V, E)$, where $V = \{1, \ldots, N\}$ denotes the set of nodes and $E \subseteq V \times V$ indicates the set of edges [23]. The adjacency matrix $Z = [z_{ij}] \in \mathbb{R}^{N \times N}$ indicates the connections among followers:

$$\begin{cases} z_{ij} = 1, & \text{if } \{j,i\} \in E \\ z_{ij} = 0, & \text{if } \{j,i\} \notin E \end{cases} \quad i,j = 1, \ldots, N, \tag{3}$$

where $\{j,i\} \in E$ means there exists a directional edge from node j to i, that is, vehicle i can receive information from j and there is no self-loop, that is, $z_{ii} = 0$. Base on Z, the Laplacian matrix $L = [l_{ij}] \in \mathbb{R}^{N \times N}$ associated with G is

$$l_{ij} = \begin{cases} \sum_{j=1}^{N} z_{ij}, & i = j \\ -z_{ij}, & i \neq j \end{cases}, \tag{4}$$

The connectivity between leader and followers is described by the pinning matrix $P = \text{diag}\{p_1, p_2, \ldots, p_N\}$, where $p_i = 1$ if node i receives information from leader, otherwise $p_i = 0$. Combined the above definitions, the topological matrix for information flow is $G = L + P$. Note that to keep the formation each follower has a direct or indirect connection with leader, which makes all eigenvalues of G be greater than zero [26].

3.3. Closed-Loop Dynamical Function of Platoon

Considering the control objective, that is, all followers track leader with predefined constant space, the following state feedback control logic with the predicted state as feedback is used [23,26,31]:

$$u_i(k) = K\{\sum_{j=1}^{N} l_{ij}[x_i(k) - \hat{x}_j(k) - d_{ij}] + p_i[x_i(k) - \hat{x}_0(k) - d_{i0}]\}, \; i = 1, \ldots, N, \tag{5}$$

where $d_{ij} = \begin{bmatrix} (i-j)d & 0 & 0 \end{bmatrix}^T$ and d is the desired clearance between neighboring vehicles, $K \in \mathbb{R}^3$ is the distributed state feedback. Furthermore, the dynamical function of control error is obtained from (1):

$$e_i(k+1) = A_i e_i(k) + B_i u_i(k) + A_i x_0(k) + B_d \rho_i(k), i = 1, \ldots, N, \tag{6}$$

where $e_i(k) = x_i(k) - x_0(k) - d_{i0}$ is the tracking control error, $B_d = \begin{bmatrix} -A_0 & -B_0 \end{bmatrix}$ and $\rho_i(k) = \begin{bmatrix} x_0(k) \\ u_0(k) \end{bmatrix}$. The uncertain closed loop dynamics of platoon control system is formulated by substituting (2) and (5) to (6):

$$\begin{aligned} E(k+1) &= (I_N \otimes A)[E(k) + X_0(k)] + (I_N \otimes H)W(k) + [G \otimes (BK)]\hat{E}(k) + (I_N \otimes B_d)\Gamma(k), \\ W(k) &= FZ(k), \; Z(k) = (I_N \otimes H_1)[E(k) + X_0(k)] + [G \otimes (H_2 K)]\hat{E}(k), \end{aligned} \tag{7}$$

where $E(k) = \begin{bmatrix} e_1(k) \\ \vdots \\ e_N(k) \end{bmatrix}, \Gamma(k) = \begin{bmatrix} \rho_1(k) \\ \vdots \\ \rho_N(k) \end{bmatrix}, \hat{E}(k) = \begin{bmatrix} \hat{x}_1(k) - \hat{x}_0(k) - d_{10} \\ \vdots \\ \hat{x}_N(k) - \hat{x}_0(k) - d_{N0} \end{bmatrix}, X_0(k) = 1_N \otimes x_0(k),$

the matrix H_1, H_2 and H are used to normalize the platoon heterogeneity to a uniform uncertain format as the following:

$$A_i = HF_iH_1, \quad B_i = HF_iH_2, \quad i = 1, \ldots, N, \\ F = \text{diag}\{F_1, \ldots, F_N\}, \|F\|_\infty \leq 1 \tag{8}$$

where F_i denotes the normalized heterogeneity of platoon.

Compared with the standard uncertain format used in robust control, (7) has the following differences: (a) An extra predicted tracking error $\hat{E}(k)$ is in the closed loop; (b) The dimension, structure and entity of G all change with running because of the degradation of communication performance, cut in/off of adjacent vehicles and so forth. To overcome these problems, a topological decoupling strategy is proposed in Section 4.

4. Synthesis of Robust Control System for Platooning

Before presenting the decoupling synthesis approach, the following lemma about matrix decomposition is introduced:

Lemma 1 ([23]). *Any matrix $P \in \mathbb{R}^{N \times N}$ has the following eigenvalue decomposition:*

$$P = X\Lambda X^{-1}, X = \tilde{X}D, \tag{9}$$

where $\Lambda = \text{diag}(\Lambda_1, \ldots, \Lambda_n) \in \mathbb{R}^{N \times N}$ is composed of the eigenvalues λ_i of P and $\sum_{i=1}^{n} \text{rank}(\Lambda_i) = N$, $\tilde{X} \in \mathbb{R}^{N \times N}$ is composed of the unit generalized eigenvectors of P, $D \in \mathbb{R}^{N \times N}$ is a diagonal matrix to convert the unit eigenvector to general one and obviously X is also composed of the generalized eigenvectors of P. The diagonal block Λ_i of Λ has four possibilities:

(a) $\Lambda_i = \lambda_i$ *if $\lambda_i \in \mathbb{R}$ and has only one linearly independent eigenvector;*

(b) *The Jordan format* $\Lambda_i = \begin{bmatrix} \lambda_i & 1 & \\ & \ddots & 1 \\ & & \lambda_i \end{bmatrix}$ *if $\lambda_i \in \mathbb{R}$ is a m-repeated eigenvalue and only has one linearly independent eigenvector;*

(c) $\Lambda_i = \begin{bmatrix} \alpha & -\beta \\ \beta & \alpha \end{bmatrix}$ *if $\lambda_i = \alpha + \beta j$ is a complex and has a pair of conjugate eigenvectors;*

(d) $\Lambda_i = \begin{bmatrix} \begin{bmatrix} \alpha & -\beta \\ \beta & \alpha \end{bmatrix} & I_2 & & \\ & \ddots & I_2 & \\ & & & \begin{bmatrix} \alpha & -\beta \\ \beta & \alpha \end{bmatrix} \end{bmatrix} \in \mathbb{R}^{2m \times 2m}$ *if $\lambda_i = \alpha + \beta j$ is a m-repeated eigenvalue and only has one pair of conjugate eigenvectors.*

4.1. Topological Decoupling of Closed-Loop Platoon System

The fundamental of decoupling synthesis strategy is depicted in Figure 3.
According to *Lemma 1*, G has the following eigenvalue decomposition:

$$G = \Theta\Lambda\Theta^{-1}, \Theta = \tilde{\Theta}D, \tag{10}$$

where $\Theta \in \mathbb{R}^{N \times N}$ is made up of the generalized eigenvectors of G, $\Lambda = \text{diag}(\Lambda_1, \ldots, \Lambda_n) \in \mathbb{R}^{N \times N}$ and $\text{rank}(\Lambda_i) = m$. Substituting (10) to (7) and the decoupled format is obtained by linear transformation as shown in Figure 3:

$$\overline{E}(k+1) = (I_N \otimes A)[\overline{E}(k) + \overline{X}_0(k)] + (I_N \otimes H)\overline{W}(k) + [\Lambda \otimes (BK)][\overline{E}(k) - \overline{\Delta}_0(k)] + \overline{\Gamma}_0(k),$$
$$\overline{Z}(k) = (I_N \otimes H_1)\overline{E}(k) + [\Lambda \otimes (H_2K)][\overline{E}(k) - \overline{\Delta}_0(k)], \quad (11)$$

where $\overline{E}(k) = (\Theta^{-1} \otimes I_3)E(k)$, $\overline{W}(k) = \overline{F}\overline{Z}(k)$, $\overline{X}_0 = (\Theta^{-1} \otimes I_3)X_0(k)$, $\overline{\Gamma}_0(k) = (\Theta^{-1} \otimes I_3)\Gamma_0(k)$, $\overline{F} = (\Theta^{-1} \otimes I_6)F(\Theta \otimes I_6)$, $\overline{\Delta}_0(k) = (\Theta^{-1} \otimes I_3)\Delta_0(k)$ and $\Delta_0(k) = 1_N u_0(k)$.

By this way, the coupling of G is partly transferred into $\overline{F}$, whose uncertain degree is related to the design of robust controller directly. For an undirected graph, G is symmetrical and $\|\overline{F}\|_\infty = \|F\|_\infty$ [11,12]. Otherwise if G is asymmetric, $\overline{F}$ may be enlarged by the linear transformation, whose bound can be estimated by:

$$\|\overline{F}\|_\infty \leq \overline{\sigma}\left(\Theta \otimes I_6\right)/\underline{\sigma}\left(\Theta \otimes I_6\right) = \overline{\sigma}(\Theta)/\underline{\sigma}(\Theta) = \rho. \quad (12)$$

Such conversion can be reduced by optimizing D with such method as singular value analysis. Base on (11), the original platoon system with higher dimension and uncertain interaction is considered as multiple independent subsystems with lower order:

$$\overline{E}_i(k+1) = [I_m \otimes A + \Lambda_i \otimes (BK)]\overline{E}_i(k) + (I_m \otimes H)\overline{W}_i(k)$$
$$- [\Lambda_i \otimes (BK)]\overline{\Delta}_{0i}(k) + [I_m \otimes B_d]\overline{\Gamma}_{0i}(k), \quad (13)$$
$$\overline{Z}_i(k) = (I_m \otimes H_1)\overline{E}_i(k) + [\Lambda_i \otimes (H_2K)][\overline{E}_i(k) - \overline{\Delta}_{0i}(k)], \quad i = 1, \ldots, n.$$

where $\overline{E}_i(k)$, $\overline{W}_i(k)$, $\overline{\Delta}_{0i}(k)$, $\overline{\Gamma}_{0i}(k)$, $\overline{Z}_i(k)$ are the corresponding parts in (11). The following theorem establishes the performance relationship between the platoon system before and after decoupling, which is the basis to numerically solve the state feedback using the decoupled format.

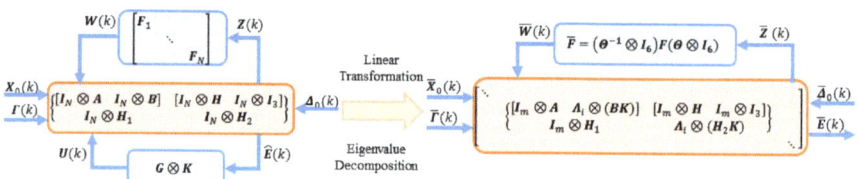

Figure 3. Fundamental of platoon decoupling.

Theorem 1. *For the platoon described by (7), if there exist positive constants $\alpha_1, \beta_1, \alpha_2, \beta_2 \in \mathbb{R}$ such that all decoupled subsystems in (13) satisfy*

$$\|\overline{Z}_i\|_2^2 \leq \alpha_1 \|\overline{W}_i\|_2^2 + \beta_1(\|\overline{\Gamma}_{0i}\|_2^2 + \|\overline{\Delta}_{0i}\|_2^2),$$
$$\|(I_m \otimes Q)\overline{E}_i\|_2^2 \leq \alpha_2 \|\overline{W}_i\|_2^2 + \beta_2(\|\overline{\Gamma}_{0i}\|_2^2 + \|\overline{\Delta}_{0i}\|_2^2), i = 1, \ldots, n. \quad (14)$$

then the original system has robust performance with the weighting matrix Q:

$$\|(I_N \otimes Q)E\|_2 \leq \mu \|\widetilde{\Gamma}\|_2, \quad (15)$$

where $\widetilde{\Gamma} = \begin{bmatrix} \Gamma \\ \Delta_0 \end{bmatrix}$ and $\mu = \rho \sqrt{\beta_2 + \frac{\alpha_2 \beta_1 \rho^2}{1 - \alpha_1 \rho^2}}$.

Proof. From (11) and the second inequality in (14), we have

$$\|(I_N \otimes Q)E\|_2^2 \leq \bar{\sigma}^2(\Theta)\left[\alpha_2\|\overline{W}\|_2^2 + \beta_2\left(\|\overline{\Gamma}\|_2^2 + \|\Delta_0\|_2^2\right)\right], \quad (16)$$
$$\|\overline{Z}\|_2^2 \leq \alpha_1\|\overline{W}_i\|_2^2 + \beta_1\left(\|\overline{\Gamma}\|_2^2 + \|\Delta_0\|_2^2\right), \quad \|\overline{W}\|_2^2 \leq \|\overline{F}\|_\infty^2 \cdot \|\overline{Z}\|_2^2 \leq \rho^2 \|\overline{Z}\|_2^2.$$

□

Substituting $\|\overline{F}\|_\infty \leq \rho$ to (16) yields (17), which is equivalent to (15) and the conclusion is proved.

$$\|(I_N \otimes Q)E\|_2 \leq \rho\sqrt{\beta_2 + \frac{\alpha_2\beta_1\rho^2}{1 - \alpha_1\rho^2}}\|\widetilde{\Gamma}\|_2. \quad (17)$$

Theorem 1 converts the performance requirements of platoon system into that of its decoupled format, whose dimension is much smaller and the unknown interactions can be evaluated by certain information, that is, eigenvalues and decomposition formats of G.

4.2. Numerical Design of State Feedback Controller

The following theorem provides a numerical way to solve the required distributed state feedback using LMI, which ensures robust stability, tracking performance and disturbance attenuation ability.

Theorem 2. *If there exist matrices $\Sigma = \Sigma^T > 0 \in \mathbb{R}^{3\times 3}$, $W \in \mathbb{R}^{1\times 3}$ and constants $\alpha_1, \alpha_2, \beta_1, \beta_2 > 0 \in \mathbb{R}$, such that the following LMIs establish:*

$$\begin{bmatrix} S_{11} & S_{12} \\ * & S_{22} \end{bmatrix} < 0, \begin{bmatrix} S_{33} & S_{34} \\ * & S_{44} \end{bmatrix} < 0, \; S_{11} = \begin{bmatrix} -I_m \otimes \Sigma & 0 \\ * & -I_{6m} \end{bmatrix},$$
$$S_{12} = \begin{bmatrix} I_m \otimes (A\Sigma) + \Lambda_i \otimes (BW) & I_m \otimes H & I_m \otimes B_d & -\Lambda_i \otimes (BW) \\ I_m \otimes (H_1\Sigma) + \Lambda_i \otimes (H_2W) & 0 & 0 & -\Lambda_i \otimes (H_2W) \end{bmatrix}$$
$$S_{22} = \mathrm{diag}(-I_m \otimes \Sigma, -\alpha_1 I_{6m}, -\beta_1 I_{12m}, -\beta_1 I_m \otimes \Sigma^2),$$
$$S_{44} = \begin{bmatrix} -\alpha_2 I_{6m} & 0 & 0 \\ * & -\beta_2 I_{12m} & 0 \\ * & * & -\beta_2 I_m \otimes \Sigma^2 \end{bmatrix}, \; S_{33} = \begin{bmatrix} -I_m \otimes \Sigma & I_m \otimes (A\Sigma) + \Lambda_i \otimes (BW) \\ * & I_m \otimes (\Sigma Q^T Q \Sigma - \Sigma) \end{bmatrix}, \quad (18)$$
$$S_{34} = \begin{bmatrix} I_m \otimes H & I_m \otimes B_d & -\Lambda_i \otimes (BW) \\ 0 & 0 & 0 \end{bmatrix}, \; i = 1, \cdots, n,$$

then with the distributed state feedback $K = W\Sigma^{-1}$, the platoon is asymptotic stable and with the weighting matrix $Q \in \mathbb{R}^{3\times 3}$, the disturbance $\widetilde{\Gamma}(k)$ is attenuated by

$$\|(I_N \otimes Q)E\|_2 \leq \mu \|\widetilde{\Gamma}\|_2. \quad (19)$$

Proof. Left and right multiplying the left side of the first inequality in (18) with $\mathrm{diag}\{I_{3m}, I_{6m}, I_m \otimes \Sigma^{-1}, I_{6m}, I_{12m}, I_m \otimes \Sigma^{-1}\}$ and substituting $K = W\Sigma^{-1}$ to it, we get:

$$\begin{bmatrix} S_{11} & \overline{S}_{12} \\ * & \overline{S}_{22} \end{bmatrix} < 0, \; \overline{S}_{12} = \begin{bmatrix} I_m \otimes A + \Lambda_i \otimes (BK) & I_m \otimes H & I_m \otimes B_d & -\Lambda_i \otimes (BK) \\ I_m \otimes H_1 + \Lambda_i \otimes (H_2K) & 0 & 0 & -\Lambda_i \otimes (H_2K) \end{bmatrix}, \quad (20)$$
$$\overline{S}_{22} = \mathrm{diag}\left(-I_m \otimes \Sigma^{-1}, -\alpha_1 I_{6m}, -\beta_1 I_{12m}, -\beta_1\beta_1 I_{3m}\right).$$

□

Applying the Schur Supplement Theorem on (20) yields [32]:

$$\overline{S}_{22} - \overline{S}_{12}^T S_{11}^{-1} \overline{S}_{12} < 0. \tag{21}$$

From (21), the following inequality establishes:

$$[I_m \otimes A + \Lambda_i \otimes (BK)]^T \left(I_m \otimes \Sigma^{-1}\right)[I_m \otimes A + \Lambda_i \otimes (BK)] < I_m \otimes \Sigma^{-1} < 0. \tag{22}$$

Therefore (13) is asymptotic stable according to the Lyapunov stability theory and the corresponding Lyapunov function is $V_i(k) = \overline{E}_i^T(k)\left(I_m \otimes \Sigma^{-1}\right)\overline{E}_i(k)$. And so the platoon system is also asymptotic stable because linear transformation will not change the stability of linear system.

Furthermore defining new Lyapunov function $V_i(k) = \overline{e}_i^T(k)\Sigma^{-1}\overline{e}_i(k), \Delta V_i(k) = V_i(k+1) - V_i(k)$ and

$$\begin{aligned} L_{1i}(k) &= \|\overline{Z}_i(k)\|_2^2 - \alpha_1 \|\overline{W}_i(k)\|_2^2 - \beta_1 (\|\overline{\Gamma}_{0i}\|_2^2 + \|\overline{\Delta}_{0i}\|_2^2), \\ L_{2i}(k) &= \|(I_m \otimes Q)\overline{E}_i\|_2^2 - \alpha_2 \|\overline{W}_i(k)\|_2^2 - \beta_2 (\|\overline{\Gamma}_{0i}\|_2^2 + \|\overline{\Delta}_{0i}\|_2^2), i = 1, \dots, n. \end{aligned} \tag{23}$$

The first equation in (23) is re-written by substituting (13) to it:

$$L_{1i}(k) = \begin{bmatrix} \overline{E}_i(k) \\ \overline{W}_i(k) \\ \overline{\Gamma}_{0i}(k) \\ \overline{\Delta}_{0i}(k) \end{bmatrix}^T \left(\overline{S}_{22} - \overline{S}_{12}^T S_{11}^{-1} \overline{S}_{12}\right) \begin{bmatrix} \overline{E}_i(k) \\ \overline{W}_i(k) \\ \overline{\Gamma}_{0i}(k) \\ \overline{\Delta}_{0i}(k) \end{bmatrix} - \Delta V_i(k), \; i = 1, \dots, n. \tag{24}$$

Then we have $L_{1i}(k) < -\Delta V_i(k)$, that is,

$$\|\overline{Z}_i(k)\|_2^2 < \alpha_1 \|\overline{W}_i(k)\|_2^2 - \beta_1 \left(\|\overline{\Gamma}_{0i}\|_2^2 + \|\overline{\Delta}_{0i}\|_2^2\right) - \Delta V_i(k), \quad i = 1, \dots, n. \tag{25}$$

The initial state is assumed to be zero and $V_i(0) = 0$ and the following inequality is derived by summarizing all sampling time from $k = 0, \cdots, \overline{N}$ together:

$$\sum_{k=0}^{\overline{N}} \|\overline{Z}_i(k)\|_2^2 < \sum_{k=0}^{\overline{N}} \left[\alpha_1 \|\overline{W}_i(k)\|_2^2 - \beta_1 \left(\|\overline{\Gamma}_{0i}\|_2^2 + \|\overline{\Delta}_{0i}\|_2^2\right)\right] - V_i(\overline{N} + 1). \tag{26}$$

Since the platoon is asymptotic stable, $\lim_{\overline{N} \to \infty} V_i(\overline{N}) = 0$ and substituting it to (26) yields

$$\|\overline{Z}_i\|_2^2 \le \alpha_1 \|\overline{W}_i\|_2^2 + \beta_1 \left(\|\overline{\Gamma}_{0i}\|_2^2 + \|\overline{\Delta}_{0i}\|_2^2\right), i = 1, \dots, n. \tag{27}$$

Similar to the analysis of (20)–(27), the following inequality also establishes:

$$\left\|\left(I_m \otimes Q\right)\overline{E}_i\right\|_2^2 \le \alpha_2 \|\overline{W}_i\|_2^2 + \beta_2 \left(\|\overline{\Gamma}_{0i}\|_2^2 + \|\overline{\Delta}_{0i}\|_2^2\right), i = 1, \dots, n. \tag{28}$$

Equation (19) is obtained from *Theorem 1* by combining (27) and (28) and *Theorem 2* is proved.

5. Closed Loop Performance Analysis

5.1. Internal Stability

This section focuses on the analysis of delay on closed loop stability. To simplify the theoretical analysis, it is assumed that the time constant of vehicle drive dynamics is the same, that is, $\tau_i = \tau, i =$

$0, 1, \ldots, N$ and there is no Jordan format in the eigenvalue decomposition of G. Then the dynamics of vehicle node becomes [11,23]:

$$x_0(k+1) = \overline{A}x_i(k) + \overline{B}u_i(k), i = 0, 1, \ldots, N. \tag{29}$$

where $\overline{A} = \begin{bmatrix} 1 & h & \tau^2\omega + \tau h \\ 0 & 1 & -\tau\omega \\ 0 & 0 & \omega+1 \end{bmatrix}$, $\overline{B} = \begin{bmatrix} \tau^2\omega - \tau h + \frac{1}{2}h^2 \\ -\tau\omega + h \\ -\omega \end{bmatrix}$, $\omega = e^{-\frac{h}{\tau}} - 1$ and h is the sample period.

Note that there are two types of stability for platoon, that is, internal stability [17] and string stability [5]. The following analysis focuses on the influence of delay on internal stability of platoon controlled by the distributed state feedback which is pre-designed in Section 4.2. The internal stability is defined as following.

Internal stability. A linear and time-invariant platoon is said to be asymptotic stable, if and only if all the eigenvalues of its discrete closed-loop system are located in the union disk of complex plane, that is, the magnitude of all eigenvalues is less than 1.

It is known from the previous studies that the internal stability criterion is not compact enough to get the parameter range directly, because Jury Criterion uses roots of equations [33]. The following Lemma gives a sufficient and necessary condition of internal stability, which is the basis to study the influence of delay and topology on platoon internal stability.

Lemma 2 ([33]). *Given a characteristic polynomial of third-order discrete system:*

$$D(z) = z^3 + c_2 z^2 + c_1 z + c_0, \tag{30}$$

where c_0, c_1 and c_2 are coefficients and the system is stable if and only if the following inequalities establish:

$$\begin{cases} D(1) > 0, \\ -D(-1) > 0, \\ 1 > |c_0|, \\ 1 - c_0^2 > |c_1 - c_2 c_0|. \end{cases} \tag{31}$$

Similar to the analysis process of heterogeneous platoon, the characteristic polynomial of studied homogeneous platoon system in this section is obtained by combing (29) with pre-designed state feedback $K = \begin{bmatrix} k_p & k_v & k_a \end{bmatrix}$ in Section 4.2:

$$D(z) = |zI - (G + \lambda\phi K)| = z^3 + c_2 z^2 + c_1 z + c_0, \tag{32}$$

where $c_0 = y^T \overline{A}_0 y + \overline{B}_0 y - 1$, $c_1 = y^T \overline{A}_1 y + \overline{B}_1 y + 3$ and $c_2 y^T \overline{A}_2 y + \overline{B}_2 y - 3$ are the coefficients whose variables are calculated by

$$\overline{A}_0 = \begin{bmatrix} \frac{k_p \lambda}{2} & -\frac{\lambda(k_p \tau + k_v)}{2} & \frac{k_p \lambda}{4} \\ * & 2\tau\lambda(k_v - k_p \tau) & -k_p \tau \lambda \\ * & * & 0 \end{bmatrix}, \overline{A}_1 = \begin{bmatrix} 0 & -\frac{\lambda(k_p \tau + k_v)}{2} & \frac{k_p \lambda}{4} \\ * & 2\tau\lambda(k_v - k_p \tau) & 0 \\ * & * & 0 \end{bmatrix},$$

$$\overline{A}_2 = \begin{bmatrix} -\frac{k_p \lambda}{2} & 0 & 0 \\ * & 0 & 0 \\ * & * & 0 \end{bmatrix}, \overline{B}_0 = \begin{bmatrix} (k_p \tau - k_v)\lambda & (-k_p \tau + k_v)\tau\lambda + k_a \lambda - 1 & -(k_p \lambda + k_v)\tau \end{bmatrix}, \tag{33}$$

$$\overline{B}_1 = \begin{bmatrix} 2(k_v - k_p \tau)\lambda & 2(k_p \tau - k_v)\tau\lambda - 2k_a\lambda + 2 & (k_p \tau + k_v)\lambda \end{bmatrix}, y = \begin{bmatrix} h & \omega & h\omega \end{bmatrix}^T,$$

$$\overline{B}_2 = \begin{bmatrix} (k_a \tau - k_v)\lambda & (-k_p \tau + k_v)\tau\lambda + k_a \lambda - 1 & 0 \end{bmatrix}.$$

The sufficient and necessary condition ensuring internal stability is derived from Lemma 2 as:

$$\begin{cases} f_1(y) = y^T(\overline{A}_0 + \overline{A}_1 + \overline{A}_2)y + (\overline{B}_0 + \overline{B}_1 + \overline{B}_2)y > 0 \\ f_2(y) = y^T(\overline{A}_0 + \overline{A}_1 + \overline{A}_2)y + (\overline{B}_0 + \overline{B}_1 + \overline{B}_2)y + 8 > 0 \\ f_3(y) = 1 - |y^T\overline{A}_0 y + \overline{B}_0 y - 1| > 0 \\ f_4(y) = 1 - c_0^2 - |c_1 - c_2 c_0| > 0 \end{cases} \quad (34)$$

5.2. Numerical Analysis

In this section the influence of topology and delay on platoon internal stability is studied numerically based on the sufficient and necessary condition (34), from which the platoon internal stability can be measured by index $f_{min} = \min_{i=1,\cdots,4} f_i(y)$. Moreover, considering the fact that linear transformation does not change the system stability and the decoupling synthesis strategy in Section 4.1, the influence of topology is expressed by its eigenvalues. During the numerical analysis, $\tau = 0.5$, $h = 0.1$ and the state feedback is $K = \begin{bmatrix} -5.75 & -5.05 & -1.03 \end{bmatrix}$ designed by Theorem 1. The numerical analysis results are shown in Figure 4.

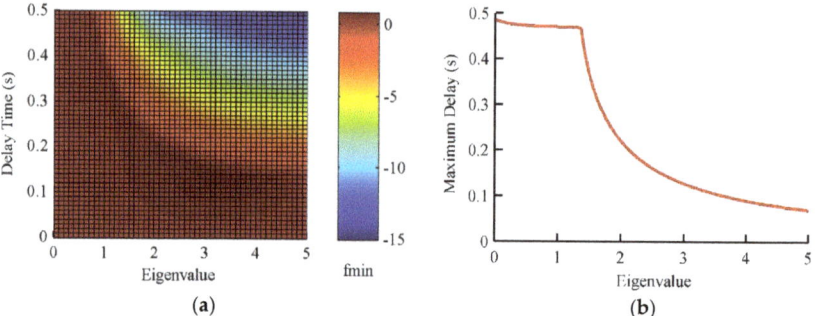

Figure 4. Influence of delay and topology on internal stability. (a) Results of f_{min}. (b) Allowable communication delay.

From (34), it is known that the platoon is internal stable if and only if $f_{min} > 0$. Then from Figure 4a, it is found that with the increasing of both of delay and eigenvalue, the platoon tends to be instable. The former phenomenon is consistent with the common sense while the latter is different from the previous conclusions without considering delay in Reference [23]. The previous studies show that both of the stability region and scalability of platoon is limited by the minimum eigenvalue of G. When considering the time delay, the internal stability is also affected by the maximum eigenvalue as shown in Figure 4b and the allowable delay decreases with the increase of it. From the decoupled format of platoon in (13), the topological eigenvalue equivalently acts on the open loop gain of control system. And from the stability theory of delay system, it is known that a higher gain is bad for the stability. According to our discovery and the communication delay is unavoidable in practical, the maximum topological eigenvalue also need to be considered when synthesizing a platoon control system.

6. Simulation and Discussion

To validate the effectiveness and further analysis of the proposed decoupling synthesis strategy, numerical simulations are conducted in this section. The simulated heterogeneous platoon includes 1 leader and 5 followers. During simulation, $v_0(0) = \cdots = v_N(0) = 5$ m/s, $d = 10$ m, $K = \begin{bmatrix} -5.75 & -5.05 & -1.03 \end{bmatrix}$, $h = 0.1$ s except where noted, the platoon heterogeneity is reflected by τ_i pre-selected in [0.3 s, 0.7 s] randomly. A statistical model is used to describe the communication delay, which is a function of the distance two communicated nodes [26]. The acceleration and velocity profiles of leader are shown in Figure 5, which are from the naturalistic driving data.

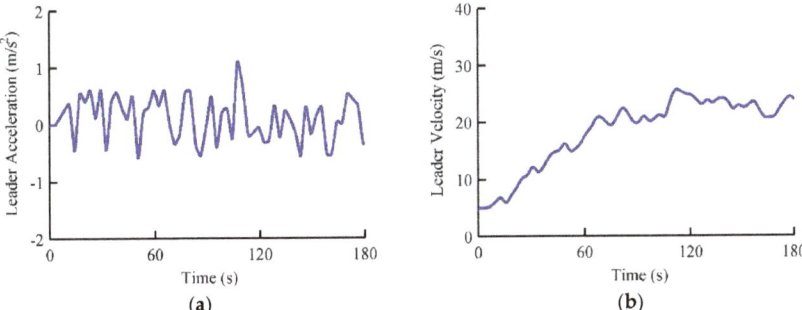

Figure 5. Profiles of leader state. (**a**) Leader Acceleration. (**b**) Leader velocity.

6.1. Delay Bound under Different Topologies

Note that the condition in *Theorem 1* is sufficient but not necessary, which implies that the designed controller still may be applicable even if LMIs in (18) are infeasible. And so in this section, we further study the actual maximum delay under different topologies. During this simulation, the maximum allowable delay is increased gradually and the maximum distance tracking error under different topologies is shown in Figure 6.

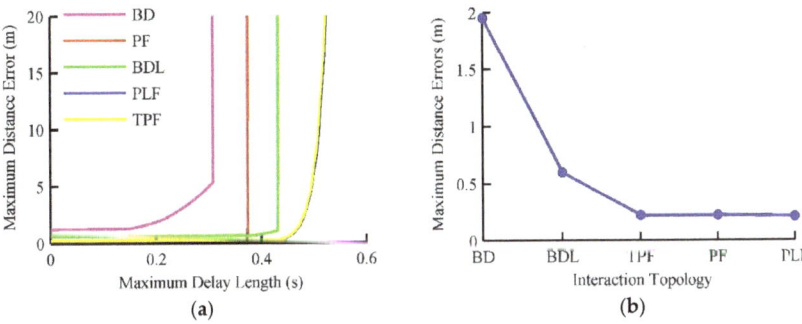

Figure 6. The upper limit of communication delay under different topologies. (**a**) Maximum delay of different topologies. (**b**) Maximum distance errors under different topologies.

As shown in Figure 6a, the maximum distance tracking error, that is, the maximum one of all followers, monotonically increases with the delay. The maximum error of BD is smaller than 2m when the delay is smaller than 0.15 s. Then it increases slowly to 5 m with the delay increases to 0.3s, after which increases sharply and even collision occurs. The maximum distance errors of other topologies agree with the trend of BD. When designing the controller, the sample period is set to 0.1s, which is large enough compared with the end-to-end communication delay of IEEE 802.11p (about 0.01s) [30]. It is found from Figure 6a that the control performance is still stable even the delay exceeds 0.1s. This shows that LMIs in (18) only are sufficient conditions and the solved controller still can control platoon acceptably even the system runs out of the designed constraints.

To show the influence of topology on distance tracking performance more clearly, the tracking error profile under the condition that the maximum delay is 0.1s is extracted shown in Figure 6b. The maximum errors of BD, BDL, TPF, PF and PLF are 1.95 m, 0.6 m, 0.22 m, 0.22 m, 0.21 m respectively.

And maximum eigenvalues of these normalized topologies are 1.9511, 1.6236, 1, 1, 1 accordingly. It agrees with the results of numerical analysis in Section 5.2 that the performance is also affected by the maximum topological eigenvalue.

6.2. Comparison of Stability Performance

To show the necessity of considering information delay, a comparison simulation between the proposed method (denoted by method 1) and an existing one (denoted by method 2) without considering delay [11] is conducted in this section. During simulation, the topology is BD having the worst performance in Section 6.1 and the maximum delay is set to 0.001s and 0.08s respectively. The former delay is negligible compared with the control period and the latter is a typical value of general V2V [30]. The compared simulation results are shown in Figure 7.

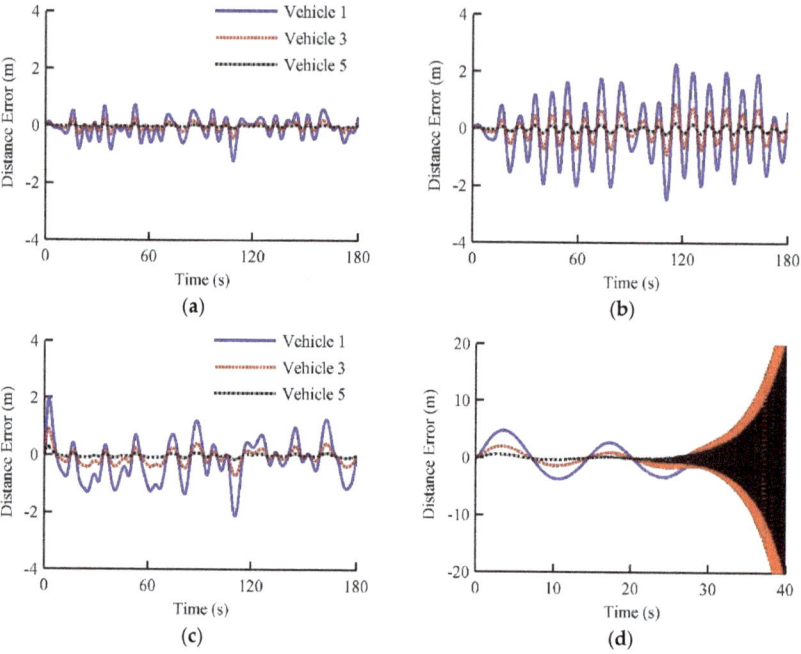

Figure 7. Compared results of different control methods. (a) Method 1 (Maximum delay is 0.001 s). (b) Method 2 (Maximum delay is 0.001 s). (c) Method 1 (Maximum delay is 0.08 s). (d) Method 2 (Maximum delay is 0.08 s).

It can be found from Figure 7a,b that the maximum distance errors of method 1 and 2 are 1.1 m and 2.5 m respectively, which both are less than the desired spacing and a stable dynamics of platoon is guaranteed when the delay is negligible. If the delay increases to 0.08s, the maximum error of method 1 is 2.2m the platoon runs stably, while the platoon controlled by method 2 becomes unstable as shown in Figure 7c,d. Moreover, the compared control results of platoon interacted by the optimal one in Section 6.1, that is, PF is shown in Figure 8. The maximum delay is set to 0.1 s and method 1 achieves a much better one than BD while method 2 still cannot ensure the platoon stability.

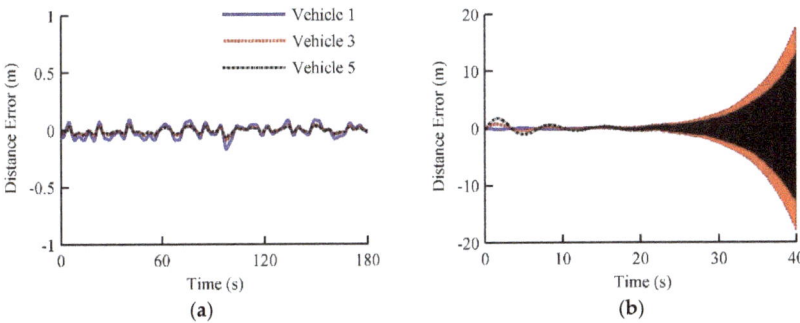

Figure 8. Stability performance of platoon interacted by PF. (**a**) Method 1 (Maximum delay is 0.1 s). (**b**) Method 2 (Maximum delay is 0.1 s).

Note that the eigenvalue varies with platoon length for some topologies such as BD. Figure 9 shows this further results of the influence of platoon length on control performance with the maximum information delay 0.08s and BD topology. As shown in Figure 9, the maximum error increases with the platoon length gradually when the vehicle number is smaller than about 140, after which the maximum error reaches about 9m and keeps almost unchanged. After the vehicle number reaches 248, the platoon become unstable. The reason is that both the minimum and maximum eigenvalues change with the dimension of BD, which affect the closed loop dynamics of platoon.

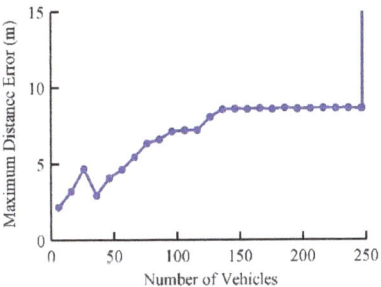

Figure 9. Influence analysis of platoon length.

6.3. Robustness Performance Analysis

The robust performance considering both delay and parametric uncertainty is studied by comparative simulations in this section to further validate the effectiveness of the proposed method. During the simulation, the platoon is interacted by BD, the platoon heterogeneity evaluated by τ_i (0 means a homogeneous platoon) and the maximum delay are increased gradually and the robust performance is measured by the maximum distance tracking error among all followers. The comparative results of robust performance are shown in Figure 10, where g_1 and g_2 represent the maximum distance errors of method 1 and method 2 respectively. When there is no information delay, the allowable disturbances of method 1 and method 2 are 2.3 and 1.8 respectively, which implies that two methods have the similar robust performance and both can attenuate the platoon heterogeneity efficiently. Furthermore, the robust performances of platoon controlled by both two methods are deteriorated with the increase of information delay but the stability region of method 1 is much wider than that of method 2 (denoted by green line). Under the condition that the maximum delay reaches 0.08 s, the allowable disturbance of method 1 is 2.2 but that of method 2 is only 0.2. This shows that method 1 has much better robustness than method 2.

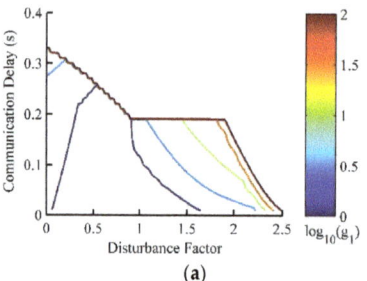 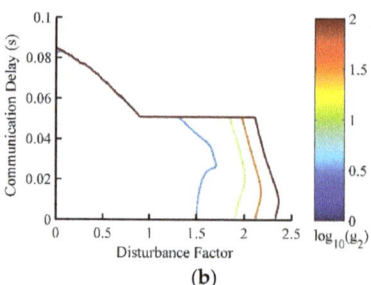

Figure 10. Robustness performance of platoon interacted by BD. (**a**) Maximum distance error of method 1. (**b**) Maximum distance error of method 2.

7. Conclusions

This paper presents a state predictor based control strategy for heterogeneous vehicular platoon connected by non-ideal wireless communication. From the theoretical analysis and simulation results, we have the following conclusions:

1. Both information delay and topological uncertainty caused by non-ideal wireless communication are critical to the stability and tracking performances of platoon, which need to be dealt with when synthesizing a platoon control system;
2. When considering the information delay, besides the minimum topological eigenvalue the maximum one also affects the closed loop performance of platoon. And comparatively, the influence of minimum one can be ignored if only stability is taken into account.
3. The proposed state predictor based control strategy can compensate for the information delay and the numerical approach based on LMI can find the required state feedback controller ensuring robust performance of platoon.

Author Contributions: B.L. and F.G. took the theoretical analysis and wrote the paper; Y.H. and C.W. performed the numerical simulation and data analysis.

Acknowledgments: This work was supported by National key research and development program under grant 2016YFB0100900 and 2016YFB0101104, Open Fund of State Key Laboratory of Vehicle NVH and Safety (NVHSKL-201705), Industrial Base Enhancement Project (2016ZXFB06002).

Conflicts of Interest: The authors declare no conflict of interest.

References

1. Banister, D.; Anderton, K.; Bonilla, D.; Givoni, M.; Schwanen, T. Transportation and the Environment. *Soc. Sci. Elect. Pub.* **2011**, *36*, 9–12. [CrossRef]
2. Li, S.; Li, R.; Wang, J.; Hu, X.; Cheng, B.; Li, K. Stabilizing Periodic Control of Automated Vehicle Platoon with Minimized Fuel Consumption. *IEEE Trans. Transp. Electr.* **2017**, *3*, 259–271. [CrossRef]
3. Swaroop, D.; Hedrick, J.K.; Choi, S.B. Direct adaptive longitudinal control of vehicle platoons. *IEEE Trans. Veh. Technol.* **2001**, *50*, 150–161. [CrossRef]
4. Kunze, R.; Ramakers, R.; Henning, K.; Jeschke, S. Organization and Operation of Electronically Coupled Truck Platoons on German Motorways. In Proceedings of the International Conference on Intelligent Robotics and Applications (ICIRA 2009), Singapore, 16–18 December 2009; pp. 135–146.
5. Shladover, S.E.; Desoer, C.A.; Hedrick, J.K.; Tomizuka, M.; Walrand, J.; Zhang, W.B.; Mckeown, N. Automated vehicle control developments in the PATH program. *IEEE Trans. Veh. Technol.* **2002**, *40*, 114–130. [CrossRef]
6. Tsugawa, S.; Kato, S.; Aoki, K. An automated truck platoon for energy saving. In Proceedings of the IEEE/RSJ International Conference on Intelligent Robots and Systems, San Francisco, CA, USA, 25–30 September 2011; pp. 4109–4114.

7. Kim, D.J.; Park, K.H.; Bien, Z. Hierarchical Longitudinal Controller for Rear-End Collision Avoidance. *IEEE Trans. Ind. Elect.* **2007**, *54*, 805–817. [CrossRef]
8. Li, S.B.; Gao, F.; Cao, D.; Li, K. Multiple-model switching control of vehicle longitudinal dynamics for platoon level automation. *IEEE Trans. Veh. Technol.* **2016**, *65*, 4480–4492. [CrossRef]
9. Caravani, P.; Santis, E.D.; Graziosi, F.; Panizzi, E. Communication Control and Driving Assistance to a Platoon of Vehicles in Heavy Traffic and Scarce Visibility. *IEEE Trans. Intell. Transp.* **2006**, *7*, 448–460. [CrossRef]
10. Liu, B.; Jia, D.; Lu, K.; Dong, N.; Wang, J.; Wu, L. A Joint Control-Communication Design for Reliable Vehicle Platooning in Hybrid Traffic. *IEEE Trans. Veh. Technol.* **2017**, *66*, 9394–9409. [CrossRef]
11. Zheng, Y.; Li, S.E.; Li, K.; Wang, L.Y. Stability Margin Improvement of Vehicular Platoon Considering Undirected Topology and Asymmetric Control. *IEEE Trans. Control Syst. Technol.* **2016**, *24*, 1253–1265. [CrossRef]
12. Xia, Q.; Gao, F.; Duan, J.; He, Y. Decoupled H-inf Control of Automated Vehicular Platoons with Complex Interaction Topologies. *IET Intell. Transp. Syst.* **2017**, *11*, 92–101. [CrossRef]
13. Rödönyi, G. An Adaptive Spacing Policy Guaranteeing String Stability in Multi-Brand Ad Hoc Platoons. *IEEE Trans. Intell. Transp.* **2017**, *19*, 1902–1912. [CrossRef]
14. Sakoda, K. Wireless communication apparatus, communication system, wireless communication apparatus control method and program. *Commun. Mag.* **2017**, *97*, 1–9.
15. Yang, N.; Wang, L.; Geraci, G.; Elkashlan, M. Safeguarding 5G wireless communication networks using physical layer security. *Commun. Magaz.* **2015**, *53*, 20–27. [CrossRef]
16. Liu, X.; Goldsmith, A.; Mahal, S.S.; Hedrick, J.K. Effects of communication delay on string stability in vehicle platoons. In Proceedings of the Intelligent Transportation Systems, Oakland, CA, USA, 25–29 August 2001; pp. 625–630.
17. Ghasemi, A.; Kazemi, R.; Azadi, S. Stable Decentralized Control of a Platoon of Vehicles with Heterogeneous Information Feedback. *IEEE Trans. Veh. Technol.* **2013**, *62*, 4299–4308. [CrossRef]
18. Alam, A.; Gattami, A.; Johansson, K.H.; Tomlin, C.J. Guaranteeing safety for heavy duty vehicle platooning: Safe set computations and experimental evaluations. *Control Eng. Pract.* **2014**, *24*, 33–41. [CrossRef]
19. Gao, F.; Hu, X.; Li, S.B.; Li, K.; Sun, Q. Distributed adaptive sliding mode control of vehicular platoon with uncertain interaction topology. *IEEE Trans. Ind. Electron.* **2018**, *65*, 6352–6361. [CrossRef]
20. Dunbar, W.B.; Caveney, D.S. Distributed Receding Horizon Control of Vehicle Platoons: Stability and String Stability. *IEEE Trans. Autom. Control* **2012**, *57*, 620–633. [CrossRef]
21. Lestas, I.; Vinnicombe, G. Scalable Robustness for Consensus Protocols with Heterogeneous Dynamics. *IFAC Proc. Vol.* **2005**, *38*, 185–190. [CrossRef]
22. Young, G.F.; Scardovi, L.; Leonard, N.E. Robustness of noisy consensus dynamics with directed communication. In Proceedings of the American Control Conference, Baltimore, MD, USA, 30 June–2 July 2010; pp. 6312–6317.
23. Zheng, Y.; Li, S.E.; Wang, J.; Cao, D.; Li, K. Stability and Scalability of Homogeneous Vehicular Platoon: Study on the Influence of Information Flow Topologies. *IEEE Trans. Intell. Transp.* **2015**, *17*, 14–26. [CrossRef]
24. Peters, A.A.; Middleton, R.H.; Mason, O. Leader tracking in homogeneous vehicle platoons with broadcast delays. *Automatica* **2014**, *50*, 64–74. [CrossRef]
25. Gong, J.; Zhao, Y.; Lu, Z. Sampled-data vehicular platoon control with communication delay. *Syst. Control Eng.* **2017**, *232*, 39–49. [CrossRef]
26. Gao, F.; Li, S.E.; Zheng, Y.; Kum, D. Robust control of heterogeneous vehicular platoon with uncertain dynamics and communication delay. *IET Intel. Transp. Syst.* **2016**, *10*, 503–513. [CrossRef]
27. Bernardo, M.D.; Salvi, A.; Santini, S. Distributed Consensus Strategy for Platooning of Vehicles in the Presence of Time-Varying Heterogeneous Communication Delays. *IEEE Trans. Intell. Transp.* **2015**, *16*, 102–112. [CrossRef]
28. Narasimha, M.; Desai, V.; Calcev, G.; Xiao, W.; Sartori, P.; Soong, A. Performance Analysis of Vehicle Platooning Using a Cellular Network. In Proceedings of the Vehicular Technology Conference, Toronto, ON, Canada, 24–27 September 2017.
29. Campolo, C.; Molinaro, A.; Araniti, G.; Berthet, A. Better Platooning Control Toward Autonomous Driving: An LTE Device-to-Device Communications Strategy That Meets Ultralow Latency Requirements. *IEEE Veh. Technol. Mag.* **2017**, *12*, 30–38. [CrossRef]

30. Masini, B.; Bazzi, A.; Zanella, A. A Survey on the Roadmap to Mandate on Board Connectivity and Enable V2V-Based Vehicular Sensor Networks. *Sensors* **2018**, *18*. [CrossRef] [PubMed]
31. Li, K.; Gao, F.; Li, S.B.; Zheng, Y.; Gao, H. Robust cooperation of connected vehicle systems with eigenvalue-bounded interaction topologies in the presence of uncertain dynamics. *Front. Mech. Eng.* **2018**, *13*, 354–367. [CrossRef]
32. Gao, F.; Li, S.E.; Kum, D.; Zhang, H. Synthesis of multiple model switching controllers using H_∞ theory for systems with large uncertainties. *Neurocomputing* **2015**, *157*, 118–124. [CrossRef]
33. Jury, E.I. A Simplified Stability Criterion for Linear Discrete Systems. *Proc. IRE* **1962**, *50*, 1493–1500. [CrossRef]

© 2019 by the authors. Licensee MDPI, Basel, Switzerland. This article is an open access article distributed under the terms and conditions of the Creative Commons Attribution (CC BY) license (http://creativecommons.org/licenses/by/4.0/).

Article

Multi-Task Scheduling Based on Classification in Mobile Edge Computing

Xiao Zheng [1], Yuanfang Chen [2,*], Muhammad Alam [3] and Jun Guo [4]

1. School of Software, Dalian University of Technology, Dalian 116620, China
2. School of Cyberspace, Hangzhou Dianzi University, Hangzhou 310018, China
3. Instituto de Telecomunicações, Campus Universitário de Santiago, 3810-193 Aveiro, Portugal
4. Key Laboratory for Ubiquitous Network and Service Software of Liaoning Province, Dalian 116620, China
* Correspondence: chenyuanfang@hdu.edu.cn; Tel.: +86-137-3545-0984

Received: 30 June 2019; Accepted: 3 August 2019; Published: 26 August 2019

Abstract: In this paper, a dynamic multi-task scheduling prototype is proposed to improve the limited resource utilization in the vehicular networks (VNET) assisted by mobile edge computing (MEC). To ensure quality of service (QoS) and meet the growing data demands, multi-task scheduling strategies should be specially constructed by considering vehicle mobility and hardware service constraints. We investigate the rational scheduling of multiple computing tasks to minimize the VNET loss. To avoid conflicts between tasks when the vehicle moves, we regard multi-task scheduling (MTS) as a multi-objective optimization (MOO) problem, and the whole goal is to find the Pareto optimal solution. Therefore, we develop some gradient-based multi-objective optimization algorithms. Those optimization algorithms are unable to deal with large-scale task scheduling because they become unscalable as the task number and gradient dimensions increase. We therefore further investigate an upper bound of the loss of multi-objective and prove that it can be optimized in an effective way. Moreover, we also reach the conclusion that, with practical assumptions, we can produce a Pareto optimal solution by upper bound optimization. Compared with the existing methods, the experimental results show that the accuracy is significantly improved.

Keywords: quality of service (QoS); vehicular networks (VNET); Mobile edge computing (MEC); multi-task scheduling (MTS); multi-objective optimization (MOO); upper bound; Pareto optimal solution

1. Introduction

Advanced wireless broadband technology has introduced an unprecedented data traffic upgrade in the vehicular networks (VNET). This aims to improve safety and fuel economy and reduce traffic congestion in the transportation system. To cope with these challenges, the offloading of tasks to road side units (RSU) has been proposed to improve quality-of-service (QoS), although a large number of computation is carried out during difficult deadline [1,2]. Nowadays, mobile smart devices become a common tool for social networks such as entertainment, learning, and smart life [3,4]. While mobile applications continue to emerge and computing power is becoming more and more dense, due to resource constraints of mobile devices (e.g., battery life and storage capacity), the computing power of mobile devices is still limited. This makes mobile users unable to achieve the same satisfaction as desktop users [5]. A more effective way to improve the performance of mobile device programs is to offload some of their tasks to the remote cloud [6,7]. However, the cloud is usually far from the mobile device, resulting in long data transmission delays between mobile devices and unpredictable results [8,9]. This is not good for mobile device programs that respond immediately. Time is of the utmost importance to mobile users, such as augmented reality apps and mobile multi-player gaming systems. To overcome the challenges mentioned above, mobile edge computing [10,11] (MEC) enables mobile devices to access their internal applications and serve a variety of wireless access networks [12,13]. This approach also enables computing tasks from the core network to be transmitted to the edge network to reduce latency. In view of some characteristics of MEC technology, multiple types of access technologies have been allowed, so vehicles can access MEC servers through various base stations (BS), such as Wi-Fi access points (Wi-Fi APs), RSU, and evolved NodeBs (eNBs).

In this paper, we assume that each edge server has the same limited resources to handle the request of the mobile vehicle, i.e., each edge server has the same processing power and these servers are arranged at certain BS locations for mobile vehicle access. We treat multiple vehicles as multiple computing tasks. Multi computing tasks scheduling problem is analogous to multi task learning (MTL) model. Due to the sharing process that produces data, even real-world tasks that appear to be unrelated have strong dependencies. This causes the application of multiple tasks to become the inductive bias in the learning model. Therefore, a multi-task scheduling (MTS) system is a model with a set of input points (co-located positions) and a set of tasks (mobile vehicles) with a variety of targets. The general workaround to construct a cross-task inductive bias is to generate a set of hypotheses that share some parameterization between tasks. In a typical MTS system, we learn its parameters by solving an optimization problem that attempts to minimize the weighted sum of the empirical losses for each task. Notice that the linear combination form makes sense if and only if the parameter set is valid in all tasks. Namely, the weighted sum of empirical loss minimization is effective only if the tasks are non-competing. However, this rarely happens. In other words, MTS is to trade-off competing objectives and merely by linear combinations will not reach the goal.

Another objective of MTS is to find solutions that are not subject to any other. This solution is denoted as the Pareto optimal. In our work, we present Pareto optimal solution for the MTS target. Finding the Pareto optimal solution that subject to a variety of quantization restraint is referred to as multi-objective optimization (MOO). There are several algorithms to solve MOO. One of these methods is called the Multi Gradient Descent Algorithm (MGDA). MGDA optimizes on the basis of gradient descent, and verifies the points that converge to the Pareto set [14]. MGDA is ideal for multi-task in deep networks. It takes advantage of the gradient of each task and solves the optimization problem while determining updates over global parameters. However, the large-scale application of MGDA is still impractical due to two technical issues. (1) Potential optimization problems cannot be extended to high dimensional gradients better, but this naturally occurs in deep networks. (2) The algorithm needs to clarify the gradient of each task, which will increase the number of backward propagations linearly, and multiply the training time by the amount of tasks.

The contributions of this paper are as follows: (1) We propose a Frank-Wolfe-based optimization algorithm and extend it to the high-dimensional problem. (2) We give the upper bound of the MGDA, and prove this method can compute by single backward propagation without designated task gradient, which greatly reduces the computation cost. (3) We prove that, in a realistic assumption condition, we can produce Pareto optimal solution by solving upper bound. (4) We conducted an empirical evaluation of the proposed method based on two different problems. First, we used MultiMNIST to make extensive assessments of multi-digit classifications [15]. Second, we applied the proposed method to Cityscape data such as joint semantic segmentation [16]. (5) Finally, the experimental results demonstrate that our method is significantly better than all baselines.

This paper is organized as follows. Section 2 presents the system model and some specific concept definitions. Section 3 abstracts the system model into a concrete multi-objective optimization problem and further finds its Pareto optimal solution. Section 4 presents the experiments on two aspects and gives the verification results. Finally, we give the conclusion.

2. Related Work

Various research issues and solutions have been considered in the edge computing literature. The strategy of caching content on various local devices is designed to bring content closer to mobile users with device-to-device (D2D) communication in HetNet [17]. We further introduce the application of the vehicular network, which has not been fully solved in the industry. Based on the reference model recommended by the MEC Industry Specification Group [18], we consider using MEC-based VNET and a MEC server to support mobile vehicle applications. The MEC server allows mobile vehicles to access edge computing resources through different wireless access methods. In particular, due to the increase in the number of vehicles, weak infrastructure, inefficient traffic control, and the frequency and severity of road traffic accidents, increasing traffic congestion was observed. We are committed to developing advanced communication technologies and intelligent data collection technologies for vehicular networks to improve safety, increase efficiency, reduce accidents, and reduce traffic congestion in transportation systems. The proposed cost-effective model ensures real-time communication between the vehicle and the RSU.

Multitasking learning (MTL) models are usually performed through hard or soft parameter sharing. In the hard parameter sharing model, a subset of parameters are shared between tasks, while other parameters are task-specific. In the soft parameter sharing model, all parameters are task-specific, but they are constrained by Bayesian prior [19] or joint dictionary [20,21]. With the success of deep MTL in computer vision [22,23], we focus on hard parameter sharing based on gradient optimization.

Multi-objective optimization solves the problem of optimizing a set of possible contrasting objectives. Of particular relevance to our work is gradient-based multi-objective optimization developed in [14,24]. These methods use multi-target Karush-Kuhn-Tucker (KKT) conditions [25] and find ways to reduce the descent direction of all targets. This approach extends the stochastic gradient descent proposed in [26,27]. Our work applies gradient-based multi-objective optimization to multi-task scheduling problem.

3. System Model

In this section, we discuss some models and special concept definitions (Figure 1).

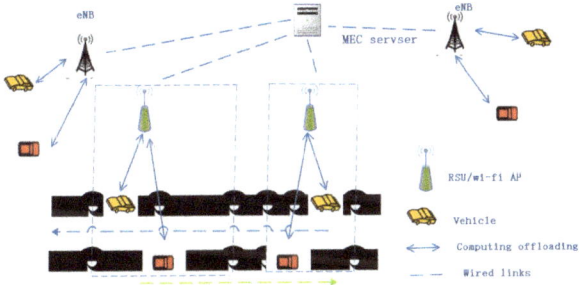

Figure 1. The structure of VNET model based on MEC.

A VNET consists of multiple eNBs, multiple RSUs that host the MEC server and multiple vehicles. The MEC server is placed around the edge of the core network rather than eNBS, which allows vehicles to access computing source via different wireless access methods. We suppose that the requesting vehicle is able to simultaneously offload the computing task and upload its tasks to the RSU by using a full-duplex technology [28,29]. We apply a MEC-based VNET to sustain multi-vehicles. Many vehicles in the vicinity of several eNBS coverage are served by the same MEC server, and expanding the range of MEC services can meet the challenges of high-speed mobile vehicles better. The eNBs that connect the entire coverage area of the MEC server are defined as the service areas of the server. To achieve resource utilization and ensure minimal VNET network loss, we consider a reasonable scheduling strategy to adjust computing tasks to different requesting vehicles and coordinate wireless access for vehicles through a wide range of resources. Here, multi-vehicles are treated as multi-task, when multiple vehicles are performing computation task offloading, there will be some impacts such as delay time and load balancing indicators on overall VNET. We need to apply the best scheduling strategy to minimize VNET loss. To avoid conflicts between multiple computing tasks when the vehicle moves, we use a multi-task scheduling (MTS) algorithm to minimize the loss of the overall network performance. Next, in the multi-task scheduling process, we model multi-task scheduling as a multi-objective optimization problem, and then try to find this Pareto optimal solution. Below, we give the specific parameters and model design.

Let $D = \{D_1, D_2, ..., D_N\}$ be the set of N computation tasks. Now, we consider the multi-task scheduling problem over a task space $\mathcal{X}$ and a set of state space $\mathcal{Y}$. To enable a large set of training sets, $\{x_i, y_i^j\}, i \in M, j \in N$ is given, where M is the number of input data points, N is the number of computing task, and y_i^j is the state of the jth task for ith input data point.

We observe a hypothesis function with respect to each task by $h^j(x; \theta', \theta^j) : \mathcal{X} \to \mathcal{Y}^j$, where parameters (θ') are shared between primary task and some special tasks (θ^j). We define the loss function for the special task as: $\mathcal{L}^j(.,.) : \mathcal{Y}^j \times \mathcal{Y}^j \to R_+$.

Although many loss functions and hypothesis functions have been mentioned in the general machine learning literature, they in general have the following objective form:

$$\min_{\theta', \theta^1, ..., \theta^N} \sum_{j=1}^{N} \alpha^j \hat{\mathcal{L}}^j(\theta', \theta^j) \tag{1}$$

where α^j is the weight of each task, and $\hat{\mathcal{L}}^j(\theta', \theta^j)$ is the empirical loss function of the task j: $\hat{\mathcal{L}}^j(\theta', \theta^j) = \frac{1}{N}\sum_i \mathcal{L}(h^j(x_i; \theta', \theta^j), y_i^j)$.

Although the weighted cumulative sum function seems attractive, it usually either obtains an expensive search or a heuristic using various scales [30,31]. The fundamental reason for scaling is the inability to define local optimality in MTS scenarios. Let us define two sets of solutions θ and $\dot{\theta}$ to enable that $\hat{\mathcal{L}}^{j_1}(\theta', \theta^{j_1}) < \hat{\mathcal{L}}^{j_1}(\dot{\theta}', \dot{\theta}^{j_1})$ and $\hat{\mathcal{L}}^{j_2}(\theta', \theta^{j_2}) > \hat{\mathcal{L}}^{j_2}(\dot{\theta}', \dot{\theta}^{j_2})$, with respect to tasks j_1 and j_2. That is, the solution of θ is better for task j_1 while $\dot{\theta}$ is favored by j_2.

On the other hand, MTS can be modeled as MOO: optimizing a set of objectives that may cause conflicts. We define the MOO function of MTS with a vector of loss function:

$$\min_{\theta', \theta^1, ..., \theta^N} L(\theta', \theta^1, ..., \theta^N) = \min_{\theta', \theta^1, ..., \theta^N} (\hat{\mathcal{L}}_1(\theta', \theta^1), ..., \hat{\mathcal{L}}_N(\theta', \theta^N))^T \tag{2}$$

The goal of MOO is to obtain the Pareto optimal solution.

The Pareto optimality of MTS is as follows:

(1) A solution θ decides a solution θ' if and only if $\hat{\mathcal{L}}^j(\theta', \theta^j) \leq \hat{\mathcal{L}}^j(\dot{\theta}', \dot{\theta}^j), \forall j$ and $L(\theta', \theta^1, ..., \theta^N) \neq L(\dot{\theta}', \dot{\theta}^1, ..., \dot{\theta}^N)$.
(2) If no solution θ dominates θ^*, then solution θ^* is called the Pareto optimal solution.

In the paper, we investigate multi-gradient descent algorithm to solve multi-objective optimization problems.

4. Multi-Objective Optimization Solution

Multi-objective optimization can solve the local optimal solution by gradient descent. In this section, we summarize a multi-gradient descent algorithm (MGDA) to solve it. MGDA makes good use of the KKT conditions necessary for optimization [24,32]. We now describe the KKT condition for the shared parameters of the primary task and the special task:

We first give the definition of KKT: the solution of general nonlinear programming problem must satisfy Karush–Kuhn–Tucker (KKT) conditions, provided that the problem constraints satisfy a regularity condition called constraint qualification. If the problem is comprised of a convex set of constraints (i.e., the solution space is convex), and the maximal value (maximal value) of objective function is concave (convex), KKT conditions are sufficient. By applying KKT condition to linear programming, we can obtain the complementary slackness conditions of primal problem and its dual problem.

4.1. Pareto-Stationarity

Let $\hat{L}^j(\theta', \theta^j)(1 \leq j \leq n, \theta', \theta \in \mathcal{B} \subseteq R^N)$ be a convex function over the open ball $\mathcal{B}$ centered at the shared parameter θ'. These functions are called Pareto optimal for the θ' point if and only if there is a convex combination of gradient vectors $u_j = \nabla_{\theta'} \hat{L}^j(\theta', \theta^j)$, in which ∇ denotes the symbol of the gradient that is equal to zero:

$$\exists c = \{c_j\} \text{ such that } c_j \geq 0 \quad \forall j, \sum_{j=1}^{n} c_j = 1, \text{ and } \sum_{j=1}^{n} c_j u_j = 0. \tag{3}$$

Thus, any solution that satisfies the condition in Equation (3) is denoted as the Pareto stable point. Each Pareto optimal is Pareto stable, and vice versa. Consider the following optimization problem:

$$\min_{c^1,\ldots,c^N} \left\{ \left\| \sum_{j=1}^{N} c_j u_j \right\|_2^2 \, \Big| \, \sum_{j=1}^{N} c_j = 1, c_j \geq 0 \quad \forall j \right\} \tag{4}$$

It means that either the solution to the optimization problem is equal to zero and the KKT condition is satisfied, or the solution solves the gradient descent direction for improving all tasks. Therefore, MTS algorithm will become a gradient reduction by solving Equation (4) on a specific task parameter and applying its solution ($\sum_{j}^{N} c_j \nabla_{\theta'}$) to the gradient update of the shared parameters. We consider how to solve Equation (4) for any model in Section 4.2 and propose an effective solution when the base model is a shared-special case in Section 4.3.

4.2. Handling the Optimal Solution

The optimal problem defined in Equation (4) is equal to seeking a minimal norm weight point in the convex hull of an input parameter space. This was originally generated in computational geometry: that is, to find a point in the convex hull that is the closest to a given query point. They have been extensively studied [33–35], and even though many researchers have proposed algorithms to solve the issue, they are not fit for our problem, because the assumptions under these algorithms are too ideal. Although some algorithms have solved the problem of finding the minimum norm in convex hulls that are composed of a large number of points of low dimensional space (usually dimension 2 or 3), under our assumption, the number of points in input parameter equals the number of tasks in the system, which is generally low. In comparison, the dimensionality, i.e., the number of shared parameters, is considered high. We therefore use a convex optimization method to solve it, because Equation (4) is also a linearly constrained convex quadratic function.

Before we solve this problem, we first consider the general situation of a two-task. The problem is formulated as $\min_{c \in [0,1]} \| c \nabla_{\theta'} \hat{L}^1(\theta', \theta^1) + (1-c) \nabla_{\theta'} \hat{L}^2(\theta', \theta^2) \|_2^2 = \min_{c \in [0,1]} \| c u_1 + (1-c) u_2 \|_2^2$.

We first show the minimum norm point in a convex hull of two input points, i.e., $\min_{\lambda \in [0,1]} \| \lambda \theta + (1-\lambda) \dot{\theta} \|_2^2$. The Algorithm is given below:

Algorithm 1: $\min_{\lambda \in [0,1]} \|\lambda \theta + (1-\lambda)\dot{\theta}\|_2^2$

1: **if** $\theta^T \dot{\theta} \geq \theta^T \theta$ **then**
2:     $\lambda = 1$
3: **else if** $\theta^T \dot{\theta} \geq \dot{\theta}^T \dot{\theta}$ **then**
4:     $\lambda = 0$
5: **else**
6:     $\lambda = \frac{(\dot{\theta}-\theta)^T \dot{\theta}}{\|\theta-\dot{\theta}\|_2^2}$
7: **end if**

Although this only applies to the case of the number of tasks $N = 2$, since the linear search can be analytically solved, this enables the Frank–Wolfe Algorithm [36] to be applied as well, thus we use it to solve the optimization constraint problem and obtain a solution:

$$\hat{c} = max\left(min\left(\left[\frac{(u_2 - u_1)^T u_2}{\|u_1 - u_2\|_2^2}\right], 1\right), 0\right) \quad (5)$$

Using the Equation (5) as a step for linear search, we obtain the update of the Frank–Wolfe Algorithm [36] as follows.

Algorithm 2: Update Algorithm for MTS

1: **for** $j = 1$ **to** N **do**
2:     $\theta^j = \theta^j - \gamma \nabla_{\theta^j} \hat{L}^j(\theta', \theta^j)$
3: **end for**
4: $c^1, ..., c^N$ = Frank–Wolfe(θ)
5: $\theta' = \theta' - \gamma \sum_{j=1}^{N} c_j u_j$
6: **Perform** Frank–Wolfe (θ) Algorithm
7: Initialize $c = (c^1, ..., c^N) = (\frac{1}{N}, ..., \frac{1}{N})$
8: Computer V s.t $V_{i,j} = (u_i)^T(u_j)$
9: **repeat**
10:    $\hat{j} = argmax_\lambda \sum_j c_j V_{\lambda j}$
11:    $\hat{\lambda} = argmin_\lambda ((1-\lambda)c + \lambda \tau_{\hat{j}})^T V((1-\lambda)c + \lambda \tau_{\hat{j}})$
12:    $c = (1-\hat{\lambda})c + \hat{\lambda}\tau_{\hat{j}}$
13: **until** $\hat{\lambda} \sim 0$ **or** Iteration Step Limit
14: **return** $c^1, ..., c^N$
15: **end perform**

4.3. Reasonable Optimization of Gradient Computation

The MTS update algorithm in Algorithm 2 is suitable for any gradient descent based optimization problem. The experimental results also show that the Frank–Wolfe algorithm is accurate and efficient because it usually converges within a moderate number of iterations and the impact on the training set is negligible. However, our proposed algorithm needs to compute u_j for every task j, a step that needs a backward propagation through shared parameters. Therefore, the resulted gradient computation will be forward propagation following N backward propagation. Since backward propagation is generally more expensive than forward propagation, the linear scaling of training time is a prohibition for many tasks.

We provide an effective way for optimizing the MTS objective and only need one post-propagation. We also show that the upper bound can produce a Pareto optimal solution under the assumption of reality. We define a shared representation function with a special task decision function that covers many existing deep multi-learning models and can also be denoted by constraining hypothesis class:

$$h^j(x; \theta', \theta^j, ..., \theta^N) = (h^j(., \theta^j)(., \theta^N) \circ r(., \theta'))(x) = h^j(r(x; \theta'); \theta^j, ..., \theta^N) \quad (6)$$

where r is the all tasks' representation function, and h^j is the specific task function that takes these representation functions as input. Now, we define the representation function as $A = (a_1, ..., a_n)$, $a_i = r(x_i, \theta')$ and we can directly express the upper bound as:

$$\left\| \sum_{j=1}^{N} c_j u_j \right\|_2^2 \leq \left\| \frac{\partial A}{\partial \theta'} \right\|_2^2 \left\| \sum_{j=1}^{N} c_j \nabla_A \hat{L}^j(\theta', \theta^j) \right\|_2^2 \quad (7)$$

where $\left\| \frac{\partial A}{\partial \theta'} \right\|_2$ is the Jacobian matrix form of A with respect to θ'. The upper bound has two properties described below: (1) $\nabla_A \hat{L}^j(\theta', \theta^j)$ is computed for all tasks in the form of a single backward propagation; and (2) $\left\| \frac{\partial A}{\partial \theta'} \right\|_2^2$ is not a function $c^1, ..., c^N$, thus it can be cancelled when it is used as an optimization objective. We use the upper bound to replace the term $\left\| \sum_{j=1}^{N} c_j u_j \right\|_2^2$. We derive the approximate optimization by throwing the $\left\| \frac{\partial A}{\partial \theta'} \right\|_2^2$ term that does not affect this optimization, so the optimization result is:

$$\min_{c^1, ..., c^N} \left\{ \left\| \sum_{j=1}^{N} c_j \nabla_A \hat{L}^j(\theta', \theta^j) \right\|_2^2 \mid \sum_{j=1}^{N} c^j = 1, c^j \geq 0 \quad \forall j \right\} \quad (8)$$

We call this problem a multi-gradient descent algorithm based on upper bound (MGDA-UB). In fact, MGDN-UB corresponds to the gradient using the task loss for the representation rather than sharing parameters. We only changed the last step of Algorithm 2.

Although the MGDA-UB is close to the original optimization algorithm, we describe it in a theorem to demonstrate the fact that our proposed algorithm yields a Pareto optimal solution under original hypothesis. Please see the proof in Appendix A.

Theorem 1. *If $c^1, ..., c^N$ is the solution of MGDA-UB and assuming $\frac{\partial A}{\partial \theta'}$ is full rank, then one of the following is true:*

(i) $\sum_{j=1}^{N} c_j u_j = 0$ *and the existing parameters are Pareto stationary.*

(ii) $\sum_{j=1}^{N} c_j u_j$ *is a descent direction which decreases every objective value.*

This conclusion results by a fact that, as long as $\frac{\partial A}{\partial \theta'}$ is full rank, the upper bound corresponding to minimizing the convex combination norm gradient defined by $\frac{\partial A}{\partial \theta'}^T \frac{\partial A}{\partial \theta'}$ using the Mahalonobis norm is optimized. This non-singular hypothesis is true because singularity means tasks are linearly related and do not need a trade-off. Overall, our proposed algorithm can find a Pareto optimal stable point with almost negligible computational cost and apply it to any MTS problem using a shared-special model.

5. Experimental Results

1. An MTS adaptation of the MNIST dataset

We evaluated the proposed MTS algorithm on the MultiMNIST dataset. Our algorithm's evaluation was based on the following three indicators as the benchmark:

(1) Uniform scaling: Minimizing the uniformly weight sum of the loss function i.e., $\frac{1}{N} \sum_j L^j$.
(2) Single task: Solving tasks individually.

(3) Grid search: Exhaustively searching for different values from $\left\{c^j \in [0,1] | \sum_j c^j = 1\right\}$ and minimizing $\frac{1}{N} \sum_j c^j L^j$.

To convert data classification into MTL problem, Sabour et al. stacked a large amount of images together. We used the same structure. For each image, we randomly selected a different image, and then placed one of these images on the upper left and the rest on the bottom right. The result is as follows: classify the data in the upper left and classify the data in the upper right. The 60 k example was used here and the existing single-task MNIST model was used directly. For the MultiMNIST experiment, we used the LeNet structure [37]. We used all layers (except last layer) as a representation function (or a shared coder). By simply adding two fully connected layers, we could treat the fully connected layer as a function of the task-special for all tasks, and each output of the representation function was taken as an input. As a loss function for a task-special, we used a softmax cross-entropy loss for both tasks. Figure 2 shows the structure.

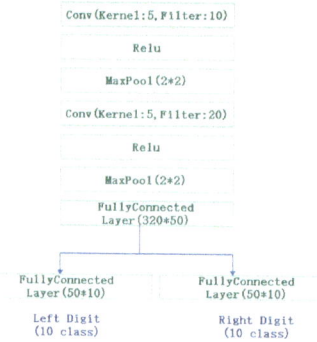

Figure 2. Structure diagram used as a MultiMNIST experiment.

We used PyTorch to do the experiment [38]. The learning rate was set as $LR = \{1 \times 10^{-3}, 5 \times 10^{-3}, 1 \times 10^{-2}, 5 \times 10^{-2}\}$, and we selected the model's learning rate that produced the highest validation correctness. We used momentum SGD. We halved the learning rate for every 20 generations, and we trained 50 times with the batch size 128. We describe the performance graph as a scatter plot of the accuracy of Task-L and Task-R in Figure 3. Our scatter plot describes the accuracy of detecting the left and right digits of all baselines. This grid shows the capacity of the computing task to compete for the system model. Our approach is the only one that embodies a solution that is as good as training a dedicated model for each computing task and is better at the top right. Table 1 shows the performance of the MTS algorithm on MultiMNIST. The single task basis can solve a single task independently with a dedicated model. The results also show that our method can find a solution that can produce as accurate solution as the single task solution. It also shows that the effectiveness of our method is as good at MTS as MOO.

Table 1. Performance of MTS algorithm on Multi MNIST.

Style	Left Digit Accuracy [%]	Right Digit Accuracy [%]
Single task	96.12	94.88
Uniform scaling	95.35	93.78
Ours	96.26	94.88

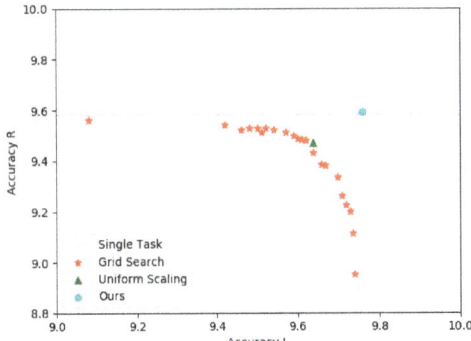

Figure 3. Diagram of MultiMNIST accuracy.

2. Scenario realization

We used GRB images to implement a more realistic scene. We dealt with three tasks: instance segmentation, semantic segmentation, and monocular depth segmentation. We used a shared-special task structure. The shared task mode was based on the full convolutional form of ResNet-50 architecture [39]. We only used the layer that is better than the average pool with full convolutions. For the special task form, we used the pyramid pooling module [40] and assigned the output scale to $256 \times 512 \times 18$ for semantic segmentation, $256 \times 512 \times 6$ for instance segmentation, and $256 \times 512 \times 2$ for monocular depth segmentation. For the loss function, we used the cross entropy with the softmax function as the semantic segmentation, the MES for the depth and instance segmentation, as shown in Figure 4.

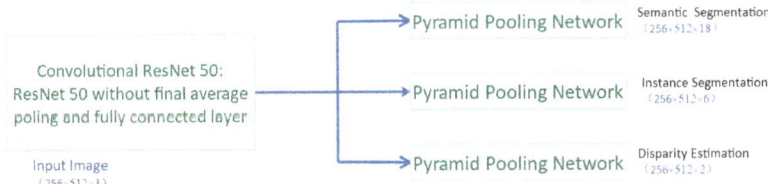

Figure 4. Scene realization experiment structure diagram.

We used pre-training on ImageNet [41] to initialize the shared task model. We applied an application of diamond pool network with bilinear interpolation that was also used by Zhou et al. [42]. Cityscapes measure set are usually not available. Thus, we used a verification set as the report result. For a validation set with hyperparametric search target, we randomly selected 260 images in the training targets. When choosing optimal hyperparameters, we used the full training target set to do retraining, kept our algorithm hidden during training and hyperparameter search, and reported measure results on Cityscapes validation set. For semantic segmentation, we measured it by mean intersection over union, and for instance segmentation we used MES. For semantic segmentation, we did not perform the cluster operation further, but reported the test results directly in the proxy task. In our experiment, we searched through the learning rate set $LR = \{1 \times 10^{-3}, 5 \times 10^{-3}, 1 \times 10^{-2}, 5 \times 10^{-2}, 1 \times 10^{-1}, 5 \times 10^{-1}\}$ and selected the model with the highest verification accuracy. We halved the learning rate for every 20 epochs, and we trained 200 times with the batch size 8. Figure 5a–c shows the performance of all baselines for semantic segmentation tasks, instance segmentation tasks, and monocular depth segmentation tasks, respectively. We used each pixel regression error to represent instance segmentation, mIoU for semantic segmentation, and disparity error for monocular depth segmentation. To convert errors of measure performance, we used 1/disparity error. We drew a two-dimensional project that represents the performance profile for each pair of computing tasks. Although we drew pairwise project for expressing visual effects, each point in the graph solves all computing tasks, and the top right is better. Table 2 shows the performance of MTS performance in three different instances. It also displays each task pair's performance, even though we performed all tasks simultaneously on three baselines.

Table 2. Performance of MTS performance in instances segmentation, semantic segmentation, and monocular depth segmentation on Cityscapes.

Style	Segmentation mIoU [%]	Instance Error [px]	Disparity Error [px]
Single task	59.67	10.25	2.68
Uniform scaling	52.36	9.32	2.81
Ours	65.76	9.12	2.45

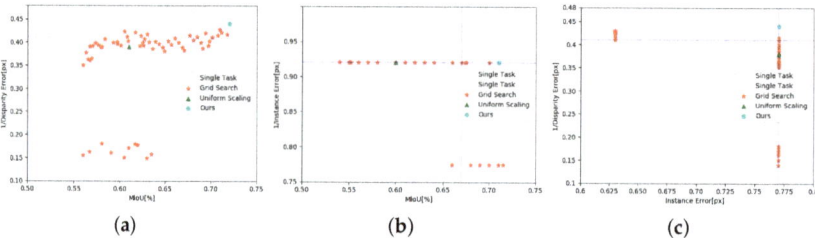

Figure 5. Scene realization experiment structure diagram.(a) shows the performance of all baselines for semantic segmentation tasks;(b) shows the performance of all baselines for instance segmentation tasks;(c) shows the performance of all baselines for monocular depth segmentation tasks.

6. Conclusions

In this paper, we form a framework of optimal resource allocation strategy for computing in vehicular networks. We formulate the optimal computing task scheduling to minimize the VNET loss under the constraints of dynamically computational resource at RSU and the limited storage capacities as well as under the constraints of hardware deadline at end-to-end and the vehicular mobility. To avoid conflicts between tasks during vehicle mobility, we convert the multi-task scheduling problem into a multi-objective optimization problem, and then find the Pareto optimal solution. For the specific large-scale vehicular network in reality, we propose a Frank–Wolfe–based MGDA optimization algorithm and extend it to the high-dimensional space. Meanwhile, we give the upper bound of the MGDA algorithm and prove it can be solved by a backward propagation without a specific-task gradient. Finally, the experimental results show that our method is greatly improved in terms of accuracy compared with the existing methods.

7. Future Works

In this paper, we study how to implement an effective and reasonable scheduling strategy in the vehicular network to minimize the performance loss of the entire network. The significance of this research is that classification-based tasks have been well promoted in the field of deep learning models, but there are certain limitations. For example, as the size of the network increases and the number of vehicles increases, there will be phenomena such as traffic congestion and insufficient cache. How to solve these problems in the super network will be considered in the future. At the same time, in the vehicular network, when the vehicle moves on the roadside base unit, privacy information will be revealed. How to design the encryption scheme will also be considered in the future.

Author Contributions: Conceptualization, methodology, and validation, X.Z., Y.C. and M.L.; Investigation and visualization, X.Z., Y.C., M.L.and J.G.; Writing—original draft preparation, X.Z., Y.C. and M.A.; Writing—review and editing, and supervision, X.Z., Y.C., J.G. and M.A.; and Funding acquisition, M.L. and Y.C.

Funding: This work was supported by the National Natural Science Foundation of China (Grant Nos. 61572095, 61877007, and 61802097), and the Project of Qianjiang Talent (Grant No. QJD1802020).

Acknowledgments: Thanks to the reviewers and editors for their careful review, constructive comments, and English editing, which helped improve the quality of the paper.

Conflicts of Interest: The authors declare no conflict of interest.

Appendix A

Proof of Theorem 1. The first case: If MGDA-UB's optimal value is 0, then that of Equation (3) is as well. We now consider the second case: if the optimal value of MGDA-UB is 0, the value of Equation (3) is not 0:

$$\sum_{j=1}^{N} c_j \nabla_{\theta'} \hat{L}^j(\theta', \theta^j) = \frac{\partial A}{\partial \theta'} \sum_{j=1}^{N} c^j \nabla_A \hat{L}^j = \sum_{j=1}^{N} c^j \nabla_{\theta'} \hat{L}^j = 0. \tag{A1}$$

Here, $c^1, ..., c^N$ is the solution of Equation (4) and the optimal value of Equation (4) is 0. This gives proof of the first case. Before we start the second case, let us give a lemma first. Since $\frac{\partial A}{\partial \theta'}$ is full rank, this equivalence is two-way. That is, if $c^1, ..., c^N$ is the solution of Equation (4), then it is also the solution of the MGDA-UB. Thus, both formulas follow Pareto stability.

We need to prove that the descent direction obtained by computing (MGDA-UB) does not increase the value of loss functions in order to prove the second case. The formal expression is given below:

$$(\sum_{j=1}^{N} c_j \nabla_{\theta'} \hat{L}^j)^T (\nabla_{\theta'} \hat{L}^{j'}) \geq 0 \quad \forall j' \in \{1, ..., N\} \tag{A2}$$

This condition is equivalent to the following:

$$(\sum_{j=1}^{N} c_j \nabla_A \hat{L}^j)^T M (\nabla_A \hat{L}^{j'}) \geq 0 \quad \forall j' \in \{1, ..., N\} \tag{A3}$$

where $M = (\frac{\partial A}{\partial \theta'})^T (\frac{\partial A}{\partial \theta'})$; since M is positive definition, this is further equivalent to the following:

$$(\sum_{j=1}^{N} c_j \nabla_A \hat{L}^j)^T (\nabla_A \hat{L}^{j'}) \geq 0 \quad \forall j' \in \{1, ..., N\} \tag{A4}$$

We prove that this is derived from the (MGDA-UB) optimality condition. The Lagrange of MGDA-UB is expressed as:

$$(\sum_{j=1}^{N} c_j \nabla_A \hat{L}^j)^T (\nabla_A \hat{L}^{j'}) - \lambda (\sum_{i} c^i - 1) \quad \text{where } \lambda \geq 0. \tag{A5}$$

Lagrange's KKT conditions yield the expected results as follows:

$$(\sum_{j=1}^{N} c_j \nabla_A \hat{L}^j)^T (\nabla_A \hat{L}^{j'}) = \frac{\lambda}{2} \geq 0. \tag{A6}$$

□

References

1. Chen, Y.; Wang, L.; Ai, Y.; Jiao, B.; Hanzo, L. Performance analysis of NOMA-SM in vehicle-to-vehicle massive MIMO channels. *IEEE J. Sel. Areas Commun.* **2017**, *35*, 2653–2666. [CrossRef]
2. Wang, J.; Jiang, C.; Han, Z.; Ren, Y.; Hanzo, L. Internet of vehicles: Sensing-aided transportation information collection and diffusion. *IEEE Trans. Veh. Technol.* **2018**, *67*, 3813–3825. [CrossRef]
3. Ahmed, E.; Gani, A.; Sookhak, M.; Ab Hamid, S.H.; Xia, F. Application optimization in mobile cloud computing: Motivation, taxonomies, and open challenges. *J. Netw. Comput. Appl.* **2015**, *52*, 52–68. [CrossRef]
4. Taleb, T.; Dutta, S.; Ksentini, A.; Iqbal, M.; Flinck, H. Mobile edge computing potential in making cities smarter. *IEEE Commun. Mag.* **2017**, *55*, 38–43. [CrossRef]
5. Dinh, H.T.; Lee, C.; Niyato, D.; Wang, P. A survey of mobile cloud computing: Architecture, applications, and approaches. *Wirel. Commun. Mob. Comput.* **2013**, *13*, 1587–1611. [CrossRef]
6. Liu, J.; Ahmed, E.; Shiraz, M.; Gani, A.; Buyya, R.; Qureshi, A. Application partitioning algorithms in mobile cloud computing: Taxonomy, review and future directions. *J. Netw. Comput. Appl.* **2015**, *48*, 99–117. [CrossRef]
7. Cloud, A.E.C. Amazon web services. *Retrieved November* **2011**, *9*, 2011.
8. Chen, Y.; Zhang, Y.; Maharjan, S.; Alam, M.; Wu, T. Deep Learning for Secure Mobile Edge Computing in Cyber-Physical Transportation Systems. *IEEE Netw.* **2019**, *33*, 36–41. [CrossRef]
9. Chen, Y.; Guizani, M.; Zhang, Y.; Wang, L.; Crespi, N.; Lee, G.M.; Wu, T. When Traffic Flow Prediction and Wireless Big Data Analytics Meet. *IEEE Netw.* **2019**, *33*, 161–167. [CrossRef]
10. Mach, P.; Becvar, Z. Mobile edge computing: A survey on architecture and computation offloading. *IEEE Commun. Surv. Tutor.* **2017**, *19*, 1628–1656. [CrossRef]
11. Shi, W.; Cao, J.; Zhang, Q.; Li, Y.; Xu, L. Edge computing: Vision and challenges. *IEEE Internet Things J.* **2016**, *3*, 637–646. [CrossRef]
12. Patel, M.; Naughton, B.; Chan, C.; Sprecher, N.; Abeta, S.; Neal, A. *Mobile-Edge Computing Introductory Technical White Paper*; White Paper, Mobile-Edge Computing (MEC) Industry Initiative; 2014. Available online: https://portal.etsi.org/portals/0/tbpages/mec/docs/mobile-edge_computing_-_introductory_technical_white_paper_v1%2018-09-14.pdf (accessed on 7 August 2019).
13. Ahmed, E.; Rehmani, M.H. Mobile edge computing: Opportunities, solutions, and challenges. *Future Gener. Comput. Syst.* **2017**, *70*, 59–63. [CrossRef]
14. Désidéri, J.A. Multiple-gradient descent algorithm (MGDA) for multiobjective optimization. *C. R. Math.* **2012**, *350*, 313–318. [CrossRef]
15. Sabour, S.; Frosst, N.; Hinton, G.E. Dynamic routing between capsules. In Proceedings of the Advances in Neural Information Processing Systems, Long Beach, NV, USA, 4–9 December 2017; pp. 3856–3866.
16. Cordts, M.; Omran, M.; Ramos, S.; Rehfeld, T.; Enzweiler, M.; Benenson, R.; Franke, U.; Roth, S.; Schiele, B. The cityscapes dataset for semantic urban scene understanding. In Proceedings of the IEEE Conference on Computer Vision and Pattern Recognition, Las Vegas, NV, USA, 26–30 June 2016; pp. 3213–3223.
17. Golrezaei, N.; Molisch, A.F.; Dimakis, A.G.; Caire, G. Femtocaching and device-to-device collaboration: A new architecture for wireless video distribution. *arXiv* **2012**, arXiv:1204.1595.
18. Mao, Y.; You, C.; Zhang, J.; Huang, K.; Letaief, K.B. A survey on mobile edge computing: The communication perspective. *IEEE Commun. Surv. Tutor.* **2017**, *19*, 2322–2358. [CrossRef]
19. Xue, Y.; Liao, X.; Carin, L.; Krishnapuram, B. Multi-task learning for classification with dirichlet process priors. *J. Mach. Learn. Res.* **2007**, *8*, 35–63.
20. Argyriou, A.; Evgeniou, T.; Pontil, M. Multi-task feature learning. In Proceedings of the 19th International Conference on Neural Information Processing Systems, Vancouver, BC, Canada, 4–7 December 2006; pp. 41–48.
21. Ruder, S. An overview of multi-task learning in deep neural networks. *arXiv* **2017**, arXiv:1706.05098.
22. Misra, I.; Shrivastava, A.; Gupta, A.; Hebert, M. Cross-stitch networks for multi-task learning. In Proceedings of the IEEE Conference on Computer Vision and Pattern Recognition, Las Vegas, NV, USA, 26–30 June 2016; pp. 3994–4003.
23. Rudd, E.M.; Günther, M.; Boult, T.E. Moon: A mixed objective optimization network for the recognition of facial attributes. In Proceedings of the European Conference on Computer Vision, Amsterdam, The Netherlands, 8–16 October 2016; pp. 19–35.

24. Schäffler, S.; Schultz, R.; Weinzierl, K. Stochastic method for the solution of unconstrained vector optimization problems. *J. Optim. Theory Appl.* **2002**, *114*, 209–222. [CrossRef]
25. Kuhn, H.W.; Tucker, A.W. Nonlinear programming. In *Proceedings of the Second Berkeley Symposium on Mathematical Statistics and Probability*; University of California Press: Berkeley, CA, USA, 31 July–12 August 1950; pp. 481–492.
26. Peitz, S.; Dellnitz, M. Gradient-based multiobjective optimization with uncertainties. In *NEO 2016*; Springer: Cham, Switzerland, 2018; pp. 159–182.
27. Poirion, F.; Mercier, Q.; Désidéri, J.A. Descent algorithm for nonsmooth stochastic multiobjective optimization. *Comput. Optim. Appl.* **2017**, *68*, 317–331. [CrossRef]
28. Zhang, Z.; Long, K.; Vasilakos, A.V.; Hanzo, L. Full-duplex wireless communications: Challenges, solutions, and future research directions. *Proc. IEEE* **2016**, *104*, 1369–1409. [CrossRef]
29. Tan, L.T.; Le, L.B. Multi-channel MAC protocol for full-duplex cognitive radio networks with optimized access control and load balancing. In Proceedings of the 2016 IEEE International Conference on Communications (ICC), Kuala Lumpur, Malaysia, 22–27 May 2016; pp. 1–6.
30. Kendall, A.; Gal, Y.; Cipolla, R. Multi-task learning using uncertainty to weigh losses for scene geometry and semantics. In Proceedings of the IEEE Conference on Computer Vision and Pattern Recognition, Salt Lake City, UT, USA, 18–22 June 2018; pp. 7482–7491.
31. Chen, Z.; Badrinarayanan, V.; Lee, C.Y.; Rabinovich, A. Gradnorm: Gradient normalization for adaptive loss balancing in deep multitask networks. *arXiv* **2017**, arXiv:1711.02257.
32. Fliege, J.; Svaiter, B.F. Steepest descent methods for multicriteria optimization. *Math. Methods Oper. Res.* **2000**, *51*, 479–494. [CrossRef]
33. Makimoto, N.; Nakagawa, I.; Tamura, A. An efficient algorithm for finding the minimum norm point in the convex hull of a finite point set in the plane. *Oper. Res. Lett.* **1994**, *16*, 33–40. [CrossRef]
34. Wolfe, P. Finding the nearest point in a polytope. *Math. Program.* **1976**, *11*, 128–149. [CrossRef]
35. Sekitani, K.; Yamamoto, Y. A recursive algorithm for finding the minimum norm point in a polytope and a pair of closest points in two polytopes. *Math. Program.* **1993**, *61*, 233–249. [CrossRef]
36. Jaggi, M. Revisiting Frank-Wolfe: Projection-Free Sparse Convex Optimization. ICML (1). 2013; pp. 427–435. Available online: http://proceedings.mlr.press/v28/jaggi13-supp.pdf (accessed on 7 August 2019)
37. LeCun, Y.; Bottou, L.; Bengio, Y.; Haffner, P. Gradient-based learning applied to document recognition. *Proc. IEEE* **1998**, *86*, 2278–2324. [CrossRef]
38. Paszke, A.; Gross, S.; Chintala, S.; Chanan, G.; Yang, E.; DeVito, Z.; Lin, Z.; Desmaison, A.; Antiga, L.; Lerer, A. Automatic Differentiation in Pytorch. NIPS Workshops 2017. Available online: https://openreview.net/forum?id=BJJsrmfCZ (accessed on 7 August 2019).
39. He, K.; Zhang, X.; Ren, S.; Sun, J. Deep residual learning for image recognition. In Proceedings of the IEEE Conference on Computer Vision and Pattern Recognition, Las Vegas, NV, USA, 26–30 June 2016; pp. 770–778.
40. Zhao, H.; Shi, J.; Qi, X.; Wang, X.; Jia, J. Pyramid scene parsing network. In Proceedings of the IEEE Conference on Computer Vision and Pattern Recognition, Honolulu, HI, USA, 21–26 July 2017; pp. 2881–2890.
41. Deng, J.; Dong, W.; Socher, R.; Li, L.J.; Li, K.; Fei-Fei, L. Imagenet: A large-scale hierarchical image database. In Proceedings of the 2009 IEEE Conference on Computer Vision and Pattern Recognition, Miami, FL, USA, 20–26 June 2009; pp. 248–255.
42. Zhou, B.; Zhao, H.; Puig, X.; Fidler, S.; Barriuso, A.; Torralba, A. Scene parsing through ade20k dataset. In Proceedings of the IEEE Conference on Computer Vision and Pattern Recognition, Honolulu, HI, USA, 21–26 July 2017; pp. 633–641.

© 2019 by the authors. Licensee MDPI, Basel, Switzerland. This article is an open access article distributed under the terms and conditions of the Creative Commons Attribution (CC BY) license (http://creativecommons.org/licenses/by/4.0/).

Article

An Adaptable Train-to-Ground Communication Architecture Based on the 5G Technological Enabler SDN

David Franco *, Marina Aguado and Nerea Toledo

Department of Communications Engineering, University of the Basque Country (UPV/EHU), 48013 Bilbao, Spain; marina.aguado@ehu.es (M.A.); nerea.toledo@ehu.eus (N.T.)
* Correspondence: david.franco@ehu.eus; Tel.: +34-946-018-251

Received: 29 April 2019; Accepted: 10 June 2019; Published: 12 June 2019

Abstract: Railway communications are closely impacted by the evolution and availability of new wireless communication technologies. Traditionally, the critical nature of railway services, the long lifecycle of rolling stock, and their certification processes challenge the adoption of the latest communication technologies. A current railway telecom trend to solve this problem is to design a flexible and adaptable communication architecture that enables the detachment of the railway services—at the application layer—and the access technologies underneath, such as 5G and beyond. One of the enablers of this detachment approach is software-defined networking (SDN)—included in 5G architecture—due to its ability to programmatically and dynamically control the network behavior via open interfaces and abstract lower-level functionalities. In this paper, we design a novel railway train-to-ground (T2G) communication architecture based on the 5G technological enabler SDN and on the transport-level redundancy technique multipath TCP (MPTCP). The goal is to provide an adaptable and multitechnology communication service while enhancing the network performance of current systems. MPTCP offers end-to-end (E2E) redundancy by the aggregation of multiple access technologies, and SDN introduces path diversity to offer a resilient and reliable communication. We carry out simulation studies to compare the performance of the legacy communication architecture with our novel approach. The results demonstrate a clear improvement in the failover response time while maintaining and even improving the uplink and downlink overall data rates.

Keywords: 5G; train-to-ground; software-defined networking; multipath TCP; adaptable; reliability; resiliency; path diversity

1. Introduction

The railway sector is ruled by a regulatory framework aiming for European interoperability. This regulatory framework trades off interoperability for adaptability to new advances in the field of communication technologies. This causes a remarkable mismatch between new railway communications services and the fast development of new wireless communication techniques.

In order to provide flexibility and adaptability to the railway network architecture [1], the application level should be independent of the access technology (2G, 5G, etc.) to permit their differently paced development. To solve this problem, the railway sector is prone to integrating a new adaptable communication layer that decouples upper-level functionalities from underlying access technologies. This concept has been named an adaptable communication system (ACS), which is understood as a set of functions and techniques that enable the aforementioned decoupling.

While solving the adaptability challenge, the envisaged communication architectures should achieve an appropriate level of reliability [2] due to the critical nature of supported services [3]. Those services have very demanding constraints that require novel techniques to assure correct system operation. These constraints are linked to short lifecycle signaling messages, which are extremely sensitive to latency, and to the continuous handover procedures, which produce high packet loss rates. Moreover, the evolution towards the IP era has introduced the challenge of providing performance indicators [4] comparable to those found in traditional dedicated systems, such as Global System for Mobile-Railway (GSM-R).

To summarize, with the purpose of achieving the reliability of legacy dedicated circuits, and complying with the aforementioned constrains, it is crucial to use resilient architectures that apply spatial and temporal redundancy techniques [5]. However, redundancy without path diversity does not provide the needed resiliency for facing correlated communication errors. This fact motivates the necessity of novel traffic engineering techniques to reserve network resources and make data travel through disjoint paths.

Figure 1 depicts the communication architecture of a traditional railway network. It consists of two parts, the access network, which connects the onboard equipment (OBE) and the control center equipment (CCE), and the multiprotocol label switching (MPLS) core, which is inspired by a real network deployed by the Italian railway infrastructure manager Rete Ferroviaria Italiana (RFI) [6].

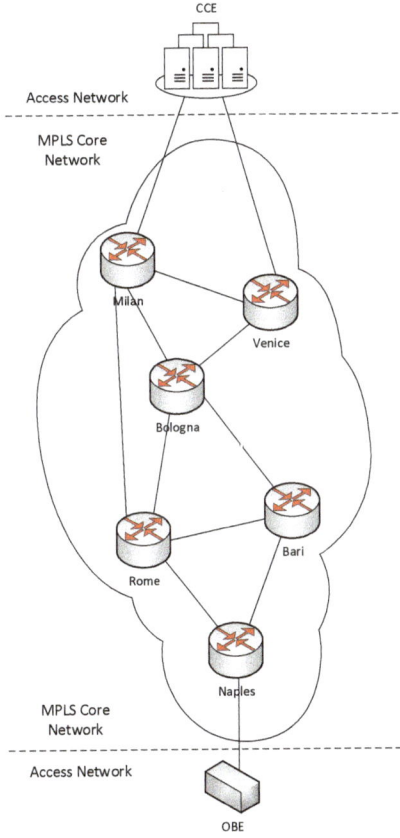

Figure 1. Legacy railway communication architecture [5]; MPLS—multiprotocol label switching, CCE—control center equipment; OBE—onboard equipment.

In contrast to this architecture, this paper proposes a detailed architecture for an ACS that is based on the 5G technological enabler SDN and on the transport-level redundancy technique multipath TCP (MPTCP) [7]. The ACS is responsible for providing resiliency mechanisms and communication interfaces that are independent of any access technology. The proposal can be divided into two functional blocks. On the one hand, end-hosts are equipped with MPTCP and multiple network interfaces in order to achieve end-to-end (E2E) spatial and temporal redundancy. This procedure reduces the latency of the communication because all the packets are duplicated and sent simultaneously, so there is no need to wait for retransmissions. On the other hand, SDN is used in the network, where a centralized SDN controller runs a novel path-computing application to make forwarding decisions. Thanks to this application, the network is capable of establishing disjoint paths and auto-reconfiguring in case of failure, which provides path diversity and improves the resiliency of the architecture. Then, the design is modeled in a simulation platform to measure the performance of this new architecture.

The structure of this document is as follows, Section 2 addresses related work in the field of adaptable and reliable mobile communication architectures. After that, Section 3 gives a detailed explanation of the proposed ACS architecture. Section 4 describes the simulation scenario that has been configured and the platform where tests have been carried out. Then, Section 5 shows the results of the simulation, and finally, Section 6 explains the main conclusions of this work.

2. Related Work

Previous works have evaluated different options to provide a reliable train-to-ground (T2G) communication through the use of resilient network architectures.

Authors in [8] divide railway user traffic into high and low priority and use relay-links to send only high priority data through redundant long-term evolution (LTE) channels. The weakness of this system lies on its lack of flexibility because data classification is static and cannot be modified on demand.

Authors in [9] propose a wireless mesh network (WMN) where the T2G communication is fully based on WiFi ad-hoc links. In this network, the train transmits data simultaneously to many trackside nodes, then packets are forwarded between nodes forming a chain until they reach the control center. This configuration improves the resiliency of the wireless channel and reduces the delay by avoiding continuous associations with the access points. Nevertheless, this solution is focused on improving the wireless access network and does not include an E2E redundant solution—core network included.

In the literature, MPTCP is also used for resiliency purposes thanks to its ability to address specific "add and drop" subflow policies to realize efficient multitechnology seamless handovers and provide E2E redundancy. The authors in [10] deploy a heterogeneous architecture using several public networks to transmit railway signaling data. However, MPTCP only guarantees the E2E redundancy and does not take care about path diversity. In [5], an MPTCP extension, which merges spatial and temporal redundancy, is proposed combined with MPLS in order to assure the required E2E path diversity and avoid correlated communication errors. However, the distributed control of MPLS does not provide enough flexibility to dynamically modify routes and reconfigure the network in case of failure. Moreover, the granularity to manage traffic flows is limited to IP source and destination directions, while when SDN features are considered, it is feasible to manage traffic flows in accordance with many other parameters.

Authors in [11] propose an architecture for T2G communication based on an SDN-controlled mobile backhaul network. This architecture can efficiently handle mobility management and provide dynamic quality-of-service (QoS) for different services onboard. There is SDN equipment placed in the ground and also inside the train to centralize the management of the wireless channel. However, they use an in-band control plane because user and control traffic is sent through the same wireless channel. This means that the failure of the wireless link will cause a communication breakdown between the SDN equipment and the controller. Consequently, the adaptability and dynamic features offered by SDN are lost.

To the best of our knowledge, our proposed architecture is the first proposal combining MPTCP and SDN that provides an adaptable, multitechnology, and reliable T2G communication service. MPTCP in both communication endpoints offers E2E redundancy while SDN in the access and core networks contributes to the path diversity and flexibility.

3. Proposed ACS Architecture

This section describes the proposed T2G network architecture. Its goal is to obtain a flexible, adaptable, and resilient communication between the train and the ground equipment in order to overcome the limitations of legacy routing. The key novelty of this architecture can be summarized in three aspects. First, to the best of our knowledge, there is no previous proposal that includes SDN and MPTCP working simultaneously. Second, MPTCP is configured to use a novel redundant scheduler, which provides spatial redundancy by duplicating every packet and sending it through different network interfaces at the same time. Third, as standalone MPTCP does not assure the use of disjoint paths, a locally customized SDN forwarding application is employed to improve path diversity. This application is able to compute and set redundant disjoint data paths—active and backup—between two given SDN switches. Moreover, the failover response time is minimized because the backup path is configured beforehand, so that the switching time is reduced to the time that takes to install a flow rule.

Furthermore, the ACS can be logically divided into three architectural components depicted in Figure 2: core network, access network, and ACS functions.

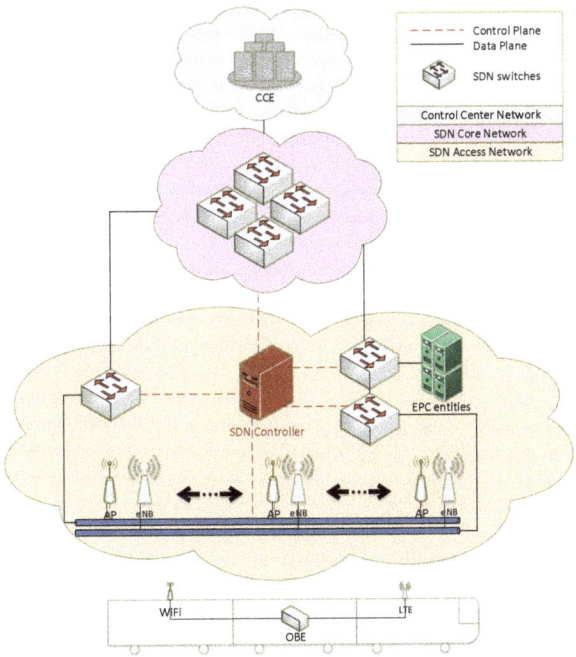

Figure 2. Proposed adaptable communication system (ACS) architecture; SDN—software-defined networking, EPC—evolved packet core, LTE—long-term evolution channel, eNB—eNodeB, AP—access point.

3.1. Core Network

In this architecture, the core network is a dedicated network managed by the railway operator. According to the SDN paradigm, there is a clear separation between control and data planes. The SDN controller is the main element of the control plane, which in this case is out-of-band because there is a point-to-point connection from the controller to each SDN-based device—OpenFlow (OF) switches and OF access points (OF-APs)—of the network (see Figure 2). Additionally, OF is set as the southbound interface, and the forwarding is based on an SDN path-computing application that is explained later.

The data plane consists of several OF switches forming a mesh topology. These switches connect the access network and the control center network, which is a private network formed by the CCE that supports railway services (e.g., telemetry).

3.2. Access Network

The access network also constitutes a dedicated network under the management of the railway operator. The control plane is the same as in the core network, except for the forwarding, which is based on L2 matching flow rules (e.g., media access control (MAC) origin–MAC destination). The data plane is divided into two subnets: WiFi and LTE. These subnets provide wireless connectivity to the OBE, which is in charge of offering T2G connectivity services to the applications running in the train.

3.2.1. WiFi

This network is fully SDN-based, having several OF-APs that are connected to an OF switch that concentrates all the traffic coming from the OF-APs. The use of additional aggregation OF switches depends on the number of deployed OF-APs.

3.2.2. LTE

This access network also supports SDN. It consists of the evolved packet core (EPC) entities and some eNodeBs (eNBs) that are connected to each other through independent OF switches. The EPC entities and the eNBs have no OF support, which means that the controller cannot modify their behavior. As in the previous case, the amount of OF switches rises according to the number of EPC entities and eNBs that are deployed. Therefore, the simplest topology would comprise two OF switches, one for the interconnection of EPC entities and the other one to aggregate the traffic from the eNBs.

3.3. ACS Functions

The ACS functionality relies on a locally developed SDN path-computing application and the MPTCP transport layer. The application runs on top of the controller, and it is focused on reducing the delay of the communication and providing path diversity. Meanwhile, MPTCP runs in the OBE and the CCE with a scheduler that uses all the available network interfaces to obtain a redundant T2G wireless communication.

3.3.1. Multipath TCP

Both sides of the communication, the OBE and the CCE, run MPTCP with the redundant scheduler [12], which was developed by our research team and added to the Linux kernel (v 3.17.0) in 2015. This scheduler provides E2E redundancy by duplicating and sending every packet through all available network interfaces, as represented by continuous red—LTE—and discontinuous green—WiFi—lines in Figure 3. The advantage of this operating mode is that the same information is sent twice, so the receiver does not need to wait for retransmissions when a packet is lost, thus the overall delay of the communication is reduced. Moreover, thanks to the redundancy offered by MPTCP, the latency of the vertical handover procedure decreases.

3.3.2. SDN Path-Computing Application

In the control plane, the SDN controller runs the Disjoint Path-Computing (DisPaC) application, which is responsible for the forwarding in the core network. DisPaC is based on the path computation element (PCE) described in [13]. The PCE calculates disjoint paths between two switches—edge switches in Figure 3—and the service manager installs the correct flow rules in the OF switches. In this case, the application has been customized to install flow rules corresponding to precomputed backup paths in order to be installed when a failure occurs.

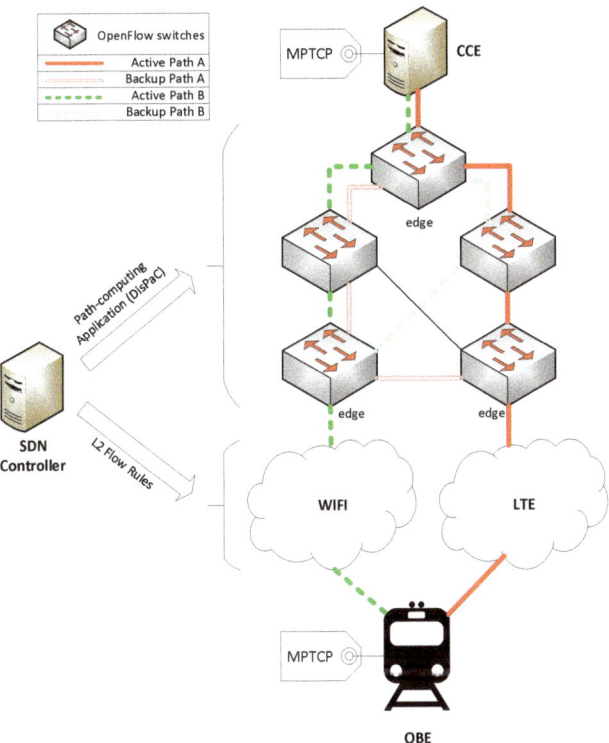

Figure 3. Data flow of the proposed ACS architecture; MPTCP—multipath TCP.

Five parameters are necessary to configure each service: source edge switch/port, destination edge switch/port and VLAN ID (to differentiate services in case of having more than one).

- If there are two disjoint paths, one path is set to active—dark color in Figure 3—and the other one to backup—light color in Figure 3. However, the flow rules corresponding to the backup path are also installed, so the traffic is also forwarded through this path. Edge switches are responsible for discarding the traffic from the backup path in order to avoid duplicated packets to be delivered out of the network. If the active path fails, the application can detect it and activate the backup path, that is, change a flow rule on each edge switch. When the active path recovers, the traffic is again handed off to it. If the backup path crashes while the active one is down, the ongoing service ends.
- If there are two different paths, but they are not fully disjoint, the application establishes them and works exactly as in the previous case.

- Finally, if there is a single path, it is established and used. When any link in the path crashes, the service stops.

In consequence, as the backup path is precomputed and ready to work, when there is a failure, the switching time between active and backup path is reduced to the time that it takes to install one flow rule, i.e., send an OF Flow-Mod message and modify the flow table. This means that the failover response time decreases and the reconfiguration is faster. Moreover, the use of disjoint paths provides path diversity, which is reflected in the resiliency of the network.

4. Simulation Scenario

This section details the simulation scenario that is configured to test the previously presented communication architecture and measure several performance indicators such as delay and data rate.

4.1. Proof of Concept Scenario

A proof of concept (PoC) scenario is designed (see Figure 4) according to the aforementioned proposed architecture.

In the control plane, the Open Network Operating System (ONOS) is used as open-source SDN controller. DisPaC is programmed over the northbound interface of ONOS, but it could be migrated to any other controller.

In the data plane, the core network consists of five OF switches forming a mesh topology. Two of them are the exchange point of WiFi and LTE access networks, and another one is directly connected to the CCE. In the access network, there are six OF-APs with a separation of 600 m between them and a single eNB located in the center between two APs, so the total trajectory is 3 km.

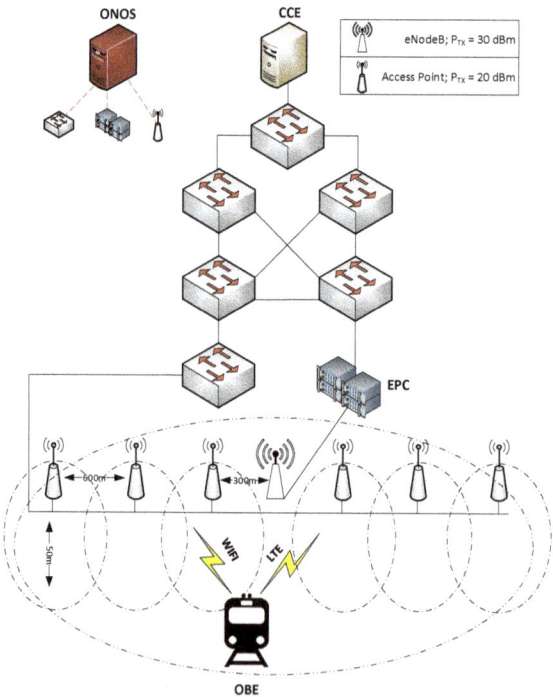

Figure 4. Proof of concept (PoC) scenario; ONOS—Open Network Operating System layout.

4.2. Simulation Framework

The tests are carried out in an open-source simulation framework called OpenNet [14] that runs on top of Mininet and ns-3. It offers a Python programming interface to create the topologies. Mininet provides SDN support by emulating OF switches that run Open vSwitch (OVS) and by enabling the connection with an external controller. In contrast, ns-3 offers WiFi and LTE channel models, as well as logical entities like OF-APs, eNBs, and the EPC (S/P-GW and MME). The ns-3 emulator counts on a Friss channel model to provide an accurate implementation of 802.11 and LTE outdoor wireless channels. The integration of SDN in WiFi and LTE networks is different. WiFi access points (APs) are fully SDN-based, and they can be managed from the controller, while SDN support in the LTE network lies in the interconnection between traditional EPC entities and eNBs. This means that those entities are treated as end-hosts by the SDN controller, so that it cannot install any flow rule to manage their operation.

4.3. Use Cases

Two use cases are considered to test different configurations of the communication network and obtain a relative comparison of delay and data rate measurements. Table 1 shows a summary of the configuration of each use case.

Table 1. Summary of use cases.

Feature	Use Case A	Use Case B
Average speed ($\bar{v}$)	72 km/h	72 km/h
Average rail traffic density	1–2 trains/h	1–2 trains/h
WiFi	IEEE 802.11g (2.4 GHz)	IEEE 802.11g (2.4 GHz)
LTE	Band 1 (2100 MHz)	Band 1 (2100 MHz)
MPTCP	✗	✓
Data forwarding method	Legacy switches with STP	SDN application (DisPaC)

The goal of these use cases is to measure the performance of the novel communication architecture. In consequence, use case A is configured according to a legacy architecture that can nowadays be found in any railway network, such as the one depicted in Figure 1. However, in order to compare equivalent mechanisms, we set a layer-2 network with switches running Spanning Tree Protocol (STP) instead of a more complex layer-3 network based on routers executing MPLS. Alternatively, the configuration of use case B represents the proposed architecture according to Section 3.

4.4. Applications

For the scope of this paper, we choose train monitoring and video surveillance applications because they cover telemetry and closed-circuit television (CCTV) services, as reported by [15].

4.4.1. Train Monitoring

In this application, the OBE periodically informs the CCE about the current state of process variables according to a publisher–subscriber paradigm. It works on top of Message Queuing Telemetry Transport (MQTT) [16], a machine-to-machine (M2M) oriented protocol that is specified for low-latency communications and which runs over TCP. In the train, the OBE collects data from the sensors and publishes it on different topics in the CCE. Each topic represents a variable, for example, the temperature of a component. Then, the CCE's subscriber entity receives all the information published on the topics of interest—the ones that it is subscribed to. This application produces an uplink overall data rate of 20 kbps per variable.

4.4.2. Video Surveillance

The video surveillance application is based on a common CCTV service, and it is modeled with the iperf3 tool. In terms of traffic characterization, it is a real-time high-quality (HQ) video stream with a resolution of 1920 × 1080 pixels (Full HD) and coded in H.265, producing an uplink data rate of around 2.5 Mbps per camera.

4.5. Measurement Techniques

Delay and data rate measurement tests are carried out over the different use cases. Delay is measured by attaching timestamps to the train monitoring messages, which makes it possible to quantify the E2E latency, including the delay introduced by transport and application levels. It is remarkable that this method supports measures over MPTCP because the MQTT protocol maintains the same TCP connection during the whole dialogue. Similarly, the iperf3 traffic generator is used to measure the instant data rate of the communication, allowing us to set up TCP flows and taking advantage of the MPTCP subsystem.

5. Results

This section presents the results obtained from the aforementioned use case simulations. The objective is to compare the performance of the proposed architecture (case B) with a legacy network (case A) when there is a failure. The failure consists in a link of the core network crashing at about 150 s after the beginning of the simulation.

Looking at Figure 5, when the failure happens, in case B there is a negligible additional latency of 1.8 ms, while in case A, the connection stops for about 50 s. In addition, Figure 6 depicts how the overall data rate increases in uplink and downlink when the SDN path-computing application (DisPaC) and MPTCP are applied. There is a 6.15% increment for the uplink data rate and 0.43% for the downlink one. It is considered that the improvement rate resulted from this comparative analysis is independent of the simulation platform and that the comparative values obtained could be extrapolated to a real scenario.

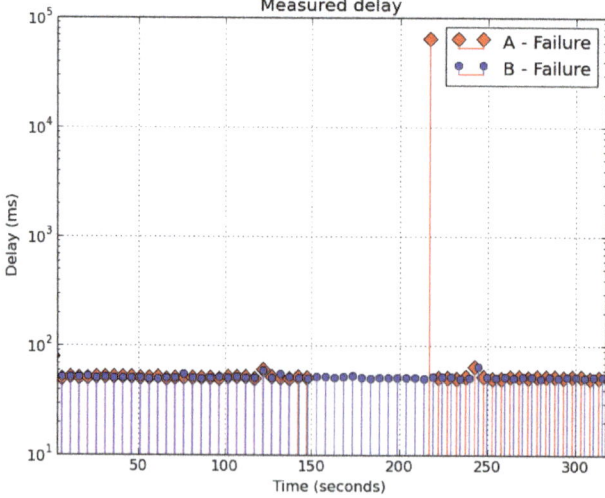

Figure 5. Delay measurement in use cases A and B when a link in the core network goes down at 150 s.

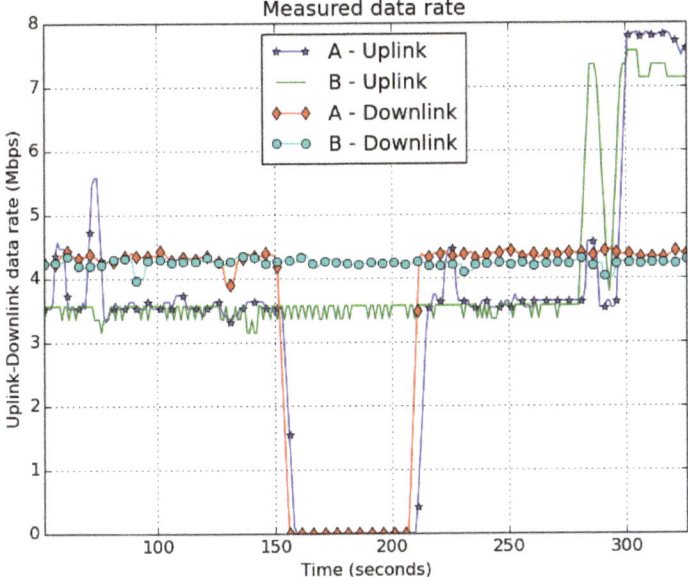

Figure 6. Data rate measurement in use cases A and B when a link in the core network goes down at 150 s.

6. Conclusions

The critical nature of railway T2G communications and the continuous evolution of radio access technologies motivates the necessity of a reliable and adaptable communication architecture that maintains the performance of legacy systems while offering enhanced functionalities for future services.

The architecture proposed in this paper relies on a 5G technological enabler, SDN, and on MPTCP to provide path diversity and E2E redundancy in order to contribute to a technology-independent and resilient communication service. SDN is a key enabler for addressing network flexibility and adaptability, due to its centralized control and its ability to deal with failures at runtime.

According to the results, the combination of MPTCP and SDN improves the T2G communication performance indicators compared to a legacy approach. The E2E latency remains in values under 60 ms (the maximum E2E delay of a user data block in GSM-R is 0.5 s [4]), the available bandwidth increments due to the provision of multiple access technologies, and the failover response time is reduced thanks to the SDN path-computing application DisPaC. DisPaC provides disjoint paths and automatic network reconfiguration with a minimal impact on performance, which improves the resiliency of the network.

Author Contributions: Conceptualization, D.F.; methodology, D.F.; software, D.F.; validation, D.F., M.A., and N.T.; formal analysis, D.F.; investigation, D.F.; data curation, D.F.; writing—original draft preparation, D.F.; writing—review and editing, M.A. and N.T.; visualization, D.F.; supervision, M.A. and N.T.; project administration, M.A. and N.T.; funding acquisition, M.A. and N.T.

Funding: This research has been supported by the Spanish Ministry of Science, Innovation, and Universities within the project TEC2017-87061-C3-1-R (CIENCIA/AEI/FEDER, UE).

Conflicts of Interest: The authors declare no conflict of interest.

Abbreviations

The following abbreviations are used in this manuscript:

5G	Fifth-generation
ACS	Adaptable communication system
AP	Access point
CCE	Control center equipment
CCTV	Closed-circuit television
DisPaC	Disjoint Path-Computing
E2E	End-to-end
eNB	evolved Node B
EPC	Evolved packet core
GSM-R	Global System for Mobile—Railway
LTE	Long-term evolution
M2M	Machine-to-machine
MAC	Media access control
MME	Mobility management entity
MPLS	Multiprotocol Label Switching
MPTCP	Multipath Transmission Control Protocol
MQTT	Message Queuing Telemetry Transport
OBE	Onboard equipment
OF	OpenFlow
ONOS	Open Network Operating System
OVS	Open vSwitch
PCE	Path computation element
QoS	Quality of service
SDN	Software-defined networking
S/P-GW	Serving/Packet-data-network Gateway
T2G	Train-to-ground
VLAN	Virtual local area network
WMN	Wireless mesh network

References

1. Banerjee, S.; Hempel, M.; Sharif, H. UPWARC: User Plane Wireless Adaptable Reliable Communication, A New Architecture for High Speed Train Passenger Internet Services. In Proceedings of the 2017 IEEE 13th International Conference on Wireless and Mobile Computing, Networking and Communications (WiMob), Rome, Italy, 9–11 October 2017; pp. 183–189. [CrossRef]
2. Wu, J.; Fan, P. A Survey on High Mobility Wireless Communications: Challenges, Opportunities and Solutions. *IEEE Access* **2016**, *4*, 450–476. [CrossRef]
3. International Union of Railways (UIC). *Future Railway Mobile Communication System: User Requirements Specification*; Technical Report, Version 3.0.0; International Union of Railways (UIC), 2018. Available online: https://uic.org/IMG/pdf/fu-7100-3.0.0.pdf (accessed on 11 June 2019).
4. European Railway Agency (ERA). *GSM-R Interfaces—Class 1 Requirements*; Informative Specification SUBSET-093 v2.3.0; European Railway Agency (ERA), 2015. Available online: https://www.era.europa.eu/ (accessed on 11 June 2019).
5. Lopez, I.; Aguado, M.; Ugarte, D.; Mendiola, A.; Higuero, M. Exploiting redundancy and path diversity for railway signalling resiliency. In Proceedings of the 2016 IEEE International Conference on Intelligent Rail Transportation (ICIRT), Birmingham, UK, 23–25 August 2016; pp. 432–439. [CrossRef]
6. UIC—Rail System Department. *IP Introduction to Railways—Version 2.0—Guideline for the Fixed Telecommunication Network*; International Union of Railways (UIC), 2013. Available online: https://www.uic.org/ (accessed on 11 June 2019).

7. Ford, A.; Raiciu, C.; Handley, M.; Bonaventure, O. TCP Extensions for Multipath Operation with Multiple Addresses. RFC 6824, RFC Editor, 2013. Available online: https://www.rfc-editor.org/info/rfc6824 (accessed on 11 June 2019).
8. Hu, F.; Zheng, K.; Long, H.; Wang, W. A cooperative hierarchical transmission scheme in railway wireless communication networks. In Proceedings of the 2011 IEEE International Conference on Service Operations, Logistics and Informatics, Beijing, China, 10–12 July 2011; pp. 605–609. [CrossRef]
9. Farooq, J.; Bro, L.; Karstensen, R.T.; Soler, J. A Multi-Radio, Multi-Hop Ad-Hoc Radio Communication Network for Communications-Based Train Control (CBTC). In Proceedings of the 2017 IEEE 86th Vehicular Technology Conference (VTC-Fall), Toronto, ON, Canada, 24–27 September 2017; pp. 1–7. [CrossRef]
10. Liu, Y.; Neri, A.; Ruggeri, A.; Vegni, A.M. A MPTCP-Based Network Architecture for Intelligent Train Control and Traffic Management Operations. *IEEE Trans. Intell. Transp. Syst.* **2017**, *18*, 2290–2302. [CrossRef]
11. Gopalasingham, A.; Van, Q.P.; Roullet, L.; Chen, C.S.; Renault, E.; Natarianni, L.; Marchi, S.D.; Hamman, E. Software-Defined mobile backhaul for future Train to ground Communication services. In Proceedings of the 2016 9th IFIP Wireless and Mobile Networking Conference (WMNC), Colmar, France, 11–13 July 2016; pp. 161–167. [CrossRef]
12. Lopez, I.; Aguado, M.; Pinedo, C.; Jacob, E. SCADA systems in the railway domain: Enhancing reliability through Redundant Multipath TCP. In Proceedings of the 2015 IEEE 18th International Conference on Intelligent Transportation Systems, Las Palmas, Spain, 15–18 September 2015. [CrossRef]
13. Mendiola, A.; Astorga, J.; Jacob, E.; Higuero, M.; Urtasun, A.; Fuentes, V. DynPaC: A Path Computation Framework for SDN. In Proceedings of the 2015 Fourth European Workshop on Software Defined Networks, Bilbao, Spain, 30 September–2 October 2015; pp. 119–120. [CrossRef]
14. Chan, M.C.; Chen, C.; Huang, J.X.; Kuo, T.; Yen, L.H.; Tseng, C.C. OpenNet: A simulator for software-defined wireless local area network. In Proceedings of the 2014 IEEE Wireless Communications and Networking Conference (WCNC), Istanbul, Turkey, 6–9 April 2014; pp. 3332–3336. [CrossRef]
15. IEC 61375-2-6. *Electronic Railway Equipment—Train Communication Network (TCN)—Part 2-6: On-Board to Ground Communication*; International Electrotechnical Commission (IEC), May 2018. Available online: https://webstore.iec.ch/publication/33737 (accessed on 11 June 2019).
16. ISO/IEC 20922. *Information Technology—Message Queuing Telemetry Transport (MQTT) v3.1.1*; International Organization for Standardization/International Electrotechnical Commission (ISO/IEC), June 2016. Available online: https://www.iso.org/standard/69466.html (accessed on 11 June 2019).

© 2019 by the authors. Licensee MDPI, Basel, Switzerland. This article is an open access article distributed under the terms and conditions of the Creative Commons Attribution (CC BY) license (http://creativecommons.org/licenses/by/4.0/).

Review

An Overview of Cooperative Driving in the European Union: Policies and Practices

Marilisa Botte [1,*], Luigi Pariota [1,2], Luca D'Acierno [1] and Gennaro Nicola Bifulco [1]

[1] Department of Civil, Architectural and Environmental Engineering, Federico II University of Naples, Via Claudio 21, 80125 Naples, Italy; luigi.pariota@unina.it (L.P.); luca.dacierno@unina.it (L.D.); gennaro.bifulco@unina.it (G.N.B.)
[2] LAERTE-ITS – Laboratory for Advanced Experiments on Roads and Traffic Environments, Federico II University of Naples, Corso Nicolangelo Protopisani 70, 80146 Naples, Italy
* Correspondence: marilisa.botte@unina.it; Tel.: +39-081-768-3356

Received: 17 April 2019; Accepted: 28 May 2019; Published: 31 May 2019

Abstract: Cooperative-Intelligent Transportation Systems (C-ITSs) aim to connect vehicles, both with one another and with road infrastructures, so as to increase traffic safety and efficiency. This paper focuses on the European framework for supporting the development of Cooperative, Connected, and Automated Mobility, and aims to shed light on the current state of testing and deployment activities in the field at the start of 2019. This may be considered particularly timely given that the year 2019 was identified as the starting date for the deployment of mature services, and the Community legislation is currently paying great attention to the matter. In order to present a concise (but comprehensive) picture, we consulted and analysed the most diverse sources comprising more than 2000 pages.

Keywords: cooperative driving; European framework; smart roads; C-ITS services; open-road pilot sites; vehicle to everything (V2X) testbeds

---

1. Introduction

Cooperative-Intelligent Transportation Systems (C-ITSs) represent the set of technological and functional elements that allow specific communication tasks identified as V2X (i.e., vehicle to everything) communication services. The 'X' in V2X can identify another vehicle (i.e., V2V communication) or the infrastructure (i.e., V2I communication). In some approaches, 'X' is also used to identify the cloud. The primary goal of such technologies is to improve road safety by helping the driver make the right decision and adapt to traffic conditions, thereby avoiding potential harm. This clearly means abating road fatalities and injury severity. Under the perspective of a progressively growing level of driving automation, these technologies could be directly exploited by the vehicle without the intervention of the driver. Other potential benefits of the use of C-ITS include enhanced traffic efficiency and improved driving experience. Indeed, the use of V2X technologies is assumed to reduce congestion and make driving tasks less reliant on human action. Finally, they can ensure environment-friendly driving through in-vehicle technologies (e.g., eco-driving), and smarter transportation management at the network level. As stated above, communication and cooperation among vehicles and between vehicles and infrastructure are crucial for the safe integration and operation of automated vehicles in transport systems of the future.

Within this framework, both the vehicles and the infrastructures have to be smart agents, and a reliable communication channel between them needs to be built. Therefore, C-ITS development has two main levers: enrichment of in-vehicle technologies and infrastructural development. Other crucial elements concern regulations, service standardisation, and cyber security issues.

Community legislation is paying great attention to the matter. In March 2019, a delegated act supplementing Directive 2010/40/EU was released [1]. Specifically, this new regulation concerning Cooperative-Intelligent Transport Systems is planned to enter into force and be directly applicable in all Member States from 31 December 2019. In light of the above, this paper aims to provide a concise (but comprehensive) overview of the current state of C-ITS testing and development activities across the European Union (EU). The challenge is to collect, examine, and summarise the countless sources available in their most diverse forms. Indeed, not only have official documents describing standards and project reports been consulted, but press releases, online articles, and datasheets have also been examined, amounting to more than 2000 pages. The goal consists in identifying C-ITS definitions, implemented technologies, level of readiness of different kinds of services (i.e., some of them are ready to be deployed while others require additional research and experimental phases), operating consortia, infrastructural development and other relevant factors, and presenting them in a synoptic and straightforward way. Clearly, although our review pays particular attention to the European case, it is worth noting that similar phenomena are occurring in the USA, Australia, and in the Far East.

The remainder of the paper is organised as follows: Section 2 describes the key features of the C-ITS European framework; Section 3 illustrates the basic principles of cooperative services and technologies; Section 4 outlines the best practices across the EU; finally, Section 5 provides the concluding remarks, including lessons learned and future perspectives.

2. The European Framework

The European Commission (EC) is making significant efforts in the field of C-ITS services, based on V2X communication, in order to set up a reference framework for supporting Cooperative, Connected, and Automated Mobility (CCAM) policies [2]. The matter is quite complex since there are several implications to be considered. Figure 1 outlines the EU framework for supporting C-ITS service development. First, C-ITS strategy aims to develop a common vision throughout the EU so as to combine the efforts of the different stakeholders involved. For this purpose, in early 2014 the EC set up the C-ITS Platform, conceived as a cooperative framework to develop a shared vision for the interoperable deployment of C-ITS in the EU. Later, in 2016, Member States and the Commission launched the C-Roads Platform to link and coordinate C-ITS deployment across the Union. Finally, on completion of the picture, there are some issues to be addressed to ensure legal certainty for the parties as well as cyber security and data privacy matters.

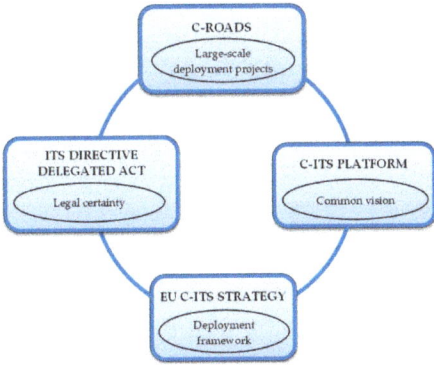

Figure 1. EU Cooperative-Intelligent Transportation Systems (C-ITS) framework.

Within the European initiative for the promotion of C-ITS, a key role is being played by the Car2Car Communication Consortium (C2C-CC) [3] and the Amsterdam Group (AG) [4].

C2C-CC it was founded in 2002 by a group of carmakers and today comprises 88 members among carmakers, Original Equipment Manufacturers (OEMs), first- and second-tier suppliers, technology suppliers, and research organisations. C2C-CC, in turn, is part of the Amsterdam Group (AG). In particular, AG is a strategic alliance formed in 2011 to support close cooperation among the stakeholders involved. Beyond C2C-CC, it comprises three other umbrella organisations: CEDR (European organisation for national road administrations), ASECAP (European association of toll road operators), and POLIS (European cities and regions network). More details about such consortia and the roadmaps they proposed for the development of cooperative driving can be found in [5].

However, it is worth pointing out that the Society of Automotive Engineers (SAE) has established a similar evolution pattern, based on different levels of increasing automation (from Level 1 to Level 5), specifically developed for autonomous driving [6]. Clearly, both for cooperative and autonomous driving, the final goal is to achieve a fully automated condition. Yet, specific correspondence between the evolution phases in the two cases has not been established, and the position of the Commission on this matter has been expressed as follows: 'even though automated vehicles do not necessarily need to be connected and connected vehicles do not require automation, it is expected that, in the medium term, connectivity will be a major enabler for driverless vehicles' [7].

To complete the picture, it is worth citing the extensive consultation process undertaken by the Commission through several working groups called upon to analyse the current situation and provide useful recommendations for the future. An appropriate instance of this is the establishment of the High Level Group (HLG) Gear 2030 in 2015. The outcome of such a consultation process is presented in [8] and provides some valuable suggestions to address the main challenges and opportunities offered by cooperative driving in the runup to 2030 and beyond.

3. Cooperative-Intelligent Transportation Systems (C-ITS) Services and Technologies Adopted in the European Commission (EC) Framework

C-ITS deployment activities have also focused on technological and functional aspects of the matter, which essentially concern the definition of a number of so-called C-ITS services and a set of reliable communication technologies.

A first macrocategory of cooperative tasks comprises *hazardous location notifications* (HLNs), which are intrinsically safety-critical services. In particular, they are aimed at providing an early warning to drivers about dangerous conditions ahead. Among them, we can find tasks for warning drivers of hard braking by vehicles ahead (i.e., *emergency electronic brake light*—EBL) as well as for providing an early warning of approaching emergency vehicles (i.e., *emergency vehicle approaching*—EVA) prior to the siren or light bar being audible or visible. Moreover, *slow or stationary Vehicle* (SSV) service is designed for warning approaching drivers about slow or stationary/broken down vehicles ahead, while *traffic jam ahead warning* (TJW) has the purpose of providing an alert to the driver on approaching the tail end of a traffic jam. This can be particularly useful when, for example, the tail end is hidden behind a hilltop or curve. Other typical HLN tasks are those concerning *road works warning* (RWW) and *weather conditions* (WTCs) services, which are aimed, respectively, at informing drivers about current road works and providing accurate and current local weather information. This is particularly useful in the case of dangerous weather conditions, which are difficult to perceive visually, such as black ice or strong gusts of wind. Additional cooperative services belonging to the HLN group are: *cooperative collision risk warning* (CCRW), whose purpose is minimising the risk of collision during overtaking or merging with traffic, and *motorcycle approaching* indication (MCA), which is finalised to warn drivers of motorcycles arriving. Finally, the *wrong-way driving* (WWD) service has been designed to provide an advance warning of wrong-way driving, thus alerting, on the one hand, drivers that they are driving in the wrong direction and, on the other, warning surrounding vehicles of the danger.

By moving towards the signage applications category we can find *in-vehicle signage* (VSGN), which is aimed at providing advance information about relevant road signs in the vehicle surroundings to increase driver awareness, and *in-vehicle speed limits* (VSPDs), whose function is to inform drivers about speed limits continuously or on a specific occurrence. Additionally, *probe vehicle data* (PVD) service is aimed at collecting vehicle data for a variety of purposes concerning circulation safety and efficiency, traffic management, and environmental issues. While, for reducing the occurrence of traffic shockwaves (i.e., transition zones between two traffic states that move through a traffic environment), thus making traffic flow smoother and reducing vehicle emissions, the *shockwave damping* (SWD) service has been developed. Moreover, *green light optimal speed advisory (GLOSA)* tasks are finalised to provide speed advice to drivers approaching traffic lights, thus reducing the number of sudden acceleration or braking incidents. In particular, for PVD, SWD, and GLOSA services, the field of application is an urban environment. Other two services belonging to signage applications are *signal violation/intersection safety* (SigV), which is a safety-critical task and aims at reducing the number and severity of collisions at signalised intersections, and *traffic signal priority request by designated vehicles* (TSP), which is aimed at allowing drivers of priority vehicles (e.g., emergency vehicles, public transport, and heavy goods vehicles (HGVs)) to be given priority at signalised junctions.

Other C-ITS services are those based on infotainment tasks whose purpose is providing different kinds of information to drivers, such as off-street and on-street parking availability, alternative routes, and location of charging stations. Further, typical services developed for urban applications are: *loading zone management* (LZM), which is conceived for supporting the management of urban parking zones specific to freight vehicles, with benefits for drivers, fleet managers and road operators; *zone access control for urban areas* (ZAC), which is targeted to provide information about access restrictions to specific urban areas, thus allowing drivers to better plan their trip; and *vulnerable road user protection* (VRU), which is a safety-critical service to the benefit of vulnerable road users, i.e., pedestrians and cyclists. Finally, it is worth citing *connected and cooperative navigation* (CCN), which represents an advanced C-ITS service for which no definitive descriptions have been released yet. However, it is based on a systemic framework in which tasks like the first and last mile, parking information, route advice, and coordinated traffic lights are fully synchronised and integrated.

A more detailed description of the features and applications of the above-mentioned C-ITS services can be found in [9–16].

Obviously, each of the services described needs to be supported by standardised messages and, as mentioned above, by suitable communication technologies. The European Telecommunications Standards Institute (ETSI) has defined two different kinds of messages, namely *cooperative awareness messages (CAMs)* [17] and *decentralized environmental notification messages (DENMs)* [18], with related packet formats and dissemination guidelines. Specifically, CAMs are a kind of heartbeat message periodically broadcast by each vehicle to its neighbours to provide information on presence, position, speed, temperature, and basic status. By contrast, DENMs are event-triggered messages broadcast to warn road users about a hazardous event. Both CAMs and DENMs are delivered to vehicles in a particular geographic region: in the immediate neighbourhood in the case of CAMs (single hop) and in the area affected by the event for DENMs (multihop) [19]. Two other kinds of standardised messages, generally disseminated by the infrastructure, are *signal phase and timing* (SPaT) and *map data (MAP)* [20–22]. The former concerns the status of traffic lights, prediction of duration and phases, data elements for prioritisation response, as well as permission linked to manoeuvres or to lanes; the latter provides information about road topology (such as topological definition of lanes for a road segment or within an intersection, links between the segments, and restrictions in lanes). To complete the picture, it is worth citing *in-vehicle information (IVI)* messages [23,24], which transmit to the vehicle static and dynamic data about infrastructure such as recommendations dictated by road signs and speed limit information.

Further, V2X data exchange can occur by means of different kinds of technologies. The most mature communication technology is described by the standard ETSI ITS G5 [25]. It is a European set of protocols and parameters for short-range communication between vehicles and traffic infrastructure, which operates in the 5.9 GHz frequency band; not coincidentally, it is also known as dedicated short-range communication (DSRC). It is based on the IEEE standard 802.11 [26] (used for Wi-Fi) and its amendment 802.11p on wireless access in vehicular environments [27], which introduces several modifications to adapt the physical (PHY) layer and medium access control (MAC) sublayer to the requirements of highly dynamic vehicular environments. More details on this technology can be found in [28].

Up and coming alternative technologies consist of cellular communications; however, this technology cannot ensure a direct link since it needs to rely on the presence of the cloud and, hence, is not suitable for use in the case of safety-critical applications [29]. Nevertheless, in this regard, clarification is due. Strictly speaking, this holds for 3G and 4G networks. Yet, with the advent of LTE (long term evolution) and fifth generation cellular communication 5G, cutting edge technologies, defined as C-V2X (cellular-V2X), have been developed. Such innovative technologies are known as LTE-V2X (or car-LTE) [30] and 5G NR-V2X (new radio-V2X) [31]. According to their advocates, they can operate equivalently to DSRC, in ITS bands and independent of cellular networks, thus ensuring a direct channel between the interlocutors involved with a very low latency [32,33]. An extensive description of LTE-V2X communication and its evolution towards fifth generation technologies can be found in [34–36]. Finally, [37–39] provide a comparative discussion about features and applications of the above communication systems together with the identification of the related optimal application ranges of distance and safety requirements.

The European Commission set in 2019 the start time of deployment of mature C-ITS services and identifies, in the C-ITS Platform Phase II Report [40], a list of services to be considered as overriding, given their high safety potential. On the basis of the priority required for their deployment, services are classified as Day-1 and Day-1.5; however, the report specifies that the above list should not be seen as the 'official C-ITS service list' but only as a first benchmark for making the various deployment activities interoperable across the EU.

Table 1 represents an excerpt from Annex 1 to the C-ITS Platform Report [40] and shows a cooperative services classification with reference to:

- interlocutors involved (i.e., V2V and V2I)
- kind of standardised messages implemented, (i.e., CAM, DENM, SPaT, MAP, and IVI)
- communication technology adopted (i.e., ETSI G5 and traditional cellular networks)
- application field (i.e., urban environment and motorway)
- safety-related features (i.e., safety-critical (SC) and nonsafety-critical (nSC))
- priority in deployment (i.e., Day-1 and Day-1.5)

Table 1. Classification of C-ITS services (source: [40]).

C-ITS Services		Interlocutors Involved			Standardised Messages				Communication Technologies		Application Field		Safety		Priority
		V2I	V2V	C	DENM	SPaT	MAP	IVI	ETSI-G5	Cellular (3G/4G)	Urban	Motorway	SC	nSC	
Hazardous Location Notifications	EBL		×		×				×		×	×	×		1
	EVA		×	×	×				×		×	×	×		1
	SSV		×	×	×				×		×	×	×		1
	TJW		×	×	×					×	×	×	×		1
	RWW	×			×					×	×	×	×		1
	WTC	×	×		×					×	×	×	×		1
	CCRW	×	×	×	×				×		×	×	×		1.5
	MCA	×	×	×	×				×		×	×	×		1.5
	WWD	×			×				×		×	×	×		1.5
Signage Applications	VSGN	×						×	×		×	×	×		1
	VSPD	×		×				×			×	×	×		1
	PVD	×		×					×		×	×	×		1
	SWD	×						×	×			×	×		1
	GLOSA	×		×		×	×		×		×		×		1
	SigV	×				×	×		×		×		×		1
	TSP	×				×	×		×		×		×		1
Others	Info [1]	×		×						×	×	×		×	1.5
	LZM	×		×						×	×			×	1.5
	ZAC	×		×						×	×			×	1.5
	VRU	V2P [2]							×		×		×		1.5
	CCN [3]	×								×				×	1.5

[1] Within infotainment tasks, the Traffic Information and Smart Routing service may also involve vehicle-to-vehicle (V2V) communication and decentralized environmental notification messages (DENMs). [2] Vulnerable road user (VRU) services are based on vehicle-to-pedestrian (V2P) communication. [3] No definitive description is given. Note: EBL, electronic brake light; EVA, emergency vehicle approaching; SSV, slow or stationary vehicle; TJW, traffic jam warning; RWW, road works warning; WTC, weather condition; CCRW, cooperative collision risk warning; MCA, motorcycle approaching; WWD, wrong way driving; VSGN, in-vehicle signage; VSPD, in-vehicle speed; PVD, probe vehicle data; SWD, shockwave damping; GLOSA, green light optimal speed advisory; SigV, signal violaion/intersection safety; TSP, traffic signal priority; LZM, loading zone management; ZAC, zone access control; CCN, connected and cooperative navigation; V2I, vehicle-to-infrastructure; CAM, cooperative awareness message; SPaT, signal phase and timing; MAP, map data; IVI, in-vehicle information; ETSI, European Telecommunications Standards Institute; SC, safety critical; and nSC, nonsafety critical.

4. Current State of the Art of C-ITS Systems

In this section, open-road and laboratory test activities for the development of cooperative driving expertise are outlined. Given the extensive nature of the topic addressed, what follows may not be considered exhaustive. Suffice it to think that connected driving accounts for around 5% of the total budget allocated for the societal challenge 'Smart Green and Integrated Transport' within the Horizon2020 research and innovation programme. As a result, there are several EU-funded projects that have been launched in the last years and many others will be initiated before the end of the seven year work programme. Instances of such projects are: AnaVANET [41], related to the development of a visualisation tool for vehicular networks; ENSEMBLE [42], related to track platooning services; MAVEN [43], related to intersection safety; 5GCar [44], related to fifth generation cellular communication; and AUTOPILOT [45], related to IoT V2X applications.

However, there are also several initiatives, such as testbeds developed by the academic community and research bodies, which, although not widely promoted or disclosed, turn out to be noteworthy and particularly relevant for taking a step forward in cooperative and automated driving tasks. First, in [46], the so-called IT-AV Automotive Environment, developed in Aveiro, is presented. It is a testbed comprising roadside units (RSUs), connected to the IT-AV datacentre through Ethernet links, and onboard units (OBUs), which are located in vehicles. They connect with each other via standard IEEE 802.11p/WAVE links and are connected to the RSUs and Internet through IEEE 802.11p/WAVE, IEEE 802.11g/Wi-Fi, or cellular links, thus enabling both V2V and V2I tasks. Therefore, on the one hand, OBUs can have access to vehicle information (such as position, heading, and speed), which are elaborated by the embedded in-car node processor to take local decisions and be notified to other vehicles. On the other hand, vehicles can also receive road information from the surroundings by means of embedded car video cameras and sensors. Such information will be transmitted to the OBUs using IEEE 802.11g/Wi-Fi. A virtualized network function (VNF) video transcoding camera-based car overtaking scenario was evaluated as a use case. However, other VNFs can also be tested such as car crash detection, emergency info dissemination, and collision avoidance.

As shown in [47], in February 2018, Telefonica and Huawei presented the 5G-V2X testbed, developed in their 5G Joint Innovation Lab in Madrid, which adopted the latest 3GPP new radio (NR) standard. In particular, a novel, self-contained frame structure for radio transmission, based on the so-called ultra-reliable and low-latency communication (URLLC) mode, was implemented for V2X applications and tested in the case of vehicle platooning applications. Additionally, to further reduce transmission latency, sparse coded multiple access (SCMA)-based grant free access technology was also investigated.

Moreover, a modular framework was adopted for the testbed described in [48], whose effectiveness was exemplarily shown by implementing a crossing assistant scenario aimed at avoiding collisions at intersections. The testbed was based on seven different modules and a three-layer plug-in software architecture. The main components were: model cars at a scale of 1:18; a specifically developed camera-based positioning system; a central control unit responsible for management, car configuration, and car control; and a visualization unit providing the status of simulation on a graphical user interface (GUI) and showing which cars are communicating as well as what information they are exchanging. In particular, the model cars are equipped with a single board computer as a processing unit, an electronic brushed motor, special hardware for attaching different sensors, and a Wi-Fi communication unit.

The main advantage of the above organisation, based on different building blocks, is that each element is easy to manage and maintain since it is independent and self-contained. Indeed, a modular approach has similarly been adopted for the development of testbeds proposed in [49] and [50]. The former is based on a software-defined radio (SDR) platform and aims to satisfy the most stringent link-level communication requirements for V2X applications, especially in the case of time-critical tasks such as emergency braking and cooperative emergency manoeuvres. The latter represents a communication-centric traffic light controller system based on the following major components: the

RSU integrated with the smart traffic light manager (STLM) module, which is responsible for the creation and transmission of SPaT and MAP messages; the OBU with the human–machine interface (HMI) displaying information in a user-friendly way; the traffic light power manager (TLPM) module, which provides the required power for traffic lights and allows them to be controlled remotely; and the traffic management curve (TMC), which monitors and manages traffic flows, thus contributing to decreased traffic jams.

A testbed specifically developed for WTC applications was presented in [51]. It was devised by the Finnish Meteorological Institute (FMI)and Lapland UAS (Lapland University of Applied Sciences), and it was tailored for typical arctic winter conditions (i.e., snow and ice). In particular, the central element was represented by a combined road weather station (RWS)/RSU hotspot, which employed IEEE 802.11p communication as the primary channel. Therefore, vehicles exchanged data with the RSU during bypass, or whenever in communication range with it, and could further transmit such information to vehicles they met outside the catchment area of the RSU. However, if IEEE 802.11p communication was not available, vehicles could obtain information also by means of traditional Wi-Fi communication or, as a last resort, a 3G cellular link to obtain current data from the nearest RWS.

A testbed based on a system configuration specifically developed for vehicle platooning applications, instead, was described in [52]. It relied, exclusively, on V2V communication by means of 802.11p and 5G-V2X vehicular technologies without the use of sensor data. Although vehicle platooning applications generally envisage that both the leader and the followers send standardised messages, in this work the communication was considered unidirectional (i.e., it occurred only from the platoon leader to the follower). It was based on CAMs, which comprised header, payload, and platooning container, while the control algorithm adopted for calculation of desired acceleration and steering angle to follow the vehicle in front was based on a model predictive controller (MPC). Moreover, application of such a service in two urban scenarios, namely fast lane change and turning manoeuvres, was presented.

Additionally, [53] presented the WiSafeCar platform for testing accident and weather condition-related services. It was based on four main components (i.e., vehicles, the RSU, the linking point, and a mobile user) and relied on an intelligent hybrid wireless traffic safety network. IEEE 802.11p communication, together with a supporting 3G cellular network, was implemented. Moreover, the possibility of exploiting vehicle-based sensor information and observation data for providing real-time services was also investigated.

Finally, it is worth citing two works that provide considerable contribution to the development of vehicular networks. The first presented a real-world testbed called HarborNet [54], a vehicular mesh networking testbed that comprised OBUs installed in trucks, administration vehicles, tow boats, and patrol vessels; RSUs connected to the optic fibre backbone of the seaport; and cloud-based data and control systems. It allowed cloud-based code deployment, remote network control, and distributed data collection from moving vehicles, thus enabling a wide range of experiments and performance analyses. HarborNet operates at the Leixões seaport in Porto, Portugal; however, it can be easily tailored to support smart city services and cooperative driving applications. By contrast, the second work [55] presented a prototyping framework as guidance for the development of simulation environments aimed at testing cooperative advanced driver-assistance system (ADAS) applications, namely cooperative emergency lane change (CELC), cooperative adaptive cruise control (CACC), and parking autonomously in cooperative environments (PACE). The proposed approach was based on an iterative prototyping process of application refining and validation, which aimed at identifying a fair compromise between simulating multiple vehicles at the same time and accuracy in modelling their dynamics.

It is worth specifying that the above-mentioned testbeds are described mainly in functional terms, while relative technical specifications can be found in the references provided.

Obviously, also beyond Europe's borders, considerable attention is paid to cooperative and autonomous driving by public authorities, road operators, equipment suppliers, and research organisations. Indeed, pilot sites and test activities are often found in the United States [56–60], China [61–64], Canada [65,66], Australia [67,68], Singapore [69,70], and Japan [71,72]. An interesting difference is that, in some of these cases, testing is more focused on the urban environment, with some emblematic cases in which the pilot zone is so extensive as to be considered a fake town. Such cases represent mock contexts where experiments may be conducted in a supervised space; the first of its kind was M-city [73], established in 2015, which covers a 130,000 m^2 area at the University of Michigan North Campus (Ann Arbor). It is a controlled simulation environment, specifically created for testing connected and autonomous driving. M-city test facilities consist, amongst other things, of various road surfaces, up to four-lane roads, roundabouts and tunnels, fixed and variable street lighting, as well as fixed and moveable buildings. Subsequently, in 2017, Uber established Almono [74], a mock city in the Hazelwood neighbourhood of Pittsburgh. It occupies 170,000 m^2 and comprises a giant roundabout, fake cars, as well as roaming mannequins and containers meant to simulate, respectively, pedestrians and buildings. However, in late 2017, the opening of an even larger pilot zone, called K-city [75], was announced by the South Korean Ministry of Land, Infrastructure, and Transport. At the moment, only motorways are operational; however, once finished, K-city will occupy 320,000 m^2. There will be several different driving conditions, including toll gates, pedestrian and train-track crossings, and even potholes and construction sites. Finally, it is worth mentioning the joint initiative of Google and its sister project Waymo (both controlled by Alphabet), aimed at testing autonomous vehicles in a fake city in California, called Castle [76], planned to reach an area of nearly 370,000 m^2.

By contrast, within Europe's borders, to the best of the authors' knowledge, two fake urban environments are currently operational, namely Zala Zone [77] in Hungary and CERMcity [78] in Germany. The former is a test environment launched in 2017 and manged by the Automotive Proving Ground Zala LTD. It is planned to cover a total area of 250 ha, where it is possible to test both classic vehicle dynamics and cooperative driving functions as well as conduct validation testing for electric vehicles. CERMcity was built by RWTH Aachen University with funding from the Federal Ministry of Education and Research. It comprises intersections, straights, parking areas, pedestrian walkways and crossings, as well as a multifunctional area and, in terms of communication technology, adopts the latest cellular standards and multiple Wi-Fi connections.

4.1. C-Roads Pilot Sites

On completion of the picture, it is worth dedicating a special mention to the C-Roads Programme. It is a large-scale deployment programme on real-life equipped infrastructures across the EU, aimed at making progress in C-ITS expertise by following a learning-by-doing approach. In December 2017, C-Roads released an overview concerning pilot sites across the EU [79]. A brief description of each national case is provided below.

4.1.1. Austria

The Austrian C-Roads-Pilot builds on the core elements of the EU C-ITS Corridor project in Austria (ECo-AT–European Corridor Austrian Testbed) [80] and extends them to a motorway-based network. The services investigated are basically Day-1 applications and particular attention is paid to the ITS-G5 communication standard. The leading competent body is the Austrian motorway operator ASFINAG, which will cover a total of 300 km of motorways and will start cross-site tests during 2019. Finally, it is worth noting that Austria is one of the Member States involved in the European project C-ITS Corridor [81], together with Germany and The Netherlands.

4.1.2. Germany

C-Roads Germany comprises two pilot sites located, respectively, in Lower Saxony and Hessen, whose test activities are harmonised by the Federal Highway Research Institute (BASt). Specifically, the companies NORDSYS and OECON P&S run the Lower Saxony pilot site, while the local public road operator Hessen Mobil is responsible for the second area. The tested services are the Day-1 type, and both ITS-G5 and cellular communications are under study. According to the planned timeline, by the end of 2019, the entire ITS system in Lower Saxony with three C-ITS services, namely SSV, VSGN, and PVD, will be operational. On the other hand, in Hessen, PVD services are already operational and will be followed by RWW, SSV, and EVA applications throughout 2019. Finally, SWD, TJM, and GLOSA services will be operational in 2020. As stated above, Germany is one of the Member States involved in the European project C-ITS Corridor.

4.1.3. The Netherlands

The Dutch pilot area is situated in the south of the Netherlands. The tested applications are Day-1 type: RWW, VSGN, and GLOSA services will be implemented in a hybrid framework combining both ITS-G5 and cellular network technologies. Moreover, infotainment tasks and different use cases of the logistic service multimodal cargo transport optimization (MCTO) are planned to be tested.

The project coordinator is the Dutch Ministry of Infrastructure and Water Management, which has set up a test site project management team (TPMT) for handling test activities. The Provinces of Noord-Brabant and Utrecht are involved. The rolling-out phase is planned to last for the whole of 2019. Moreover, road shows will be performed in 2020.

Finally, the Netherlands plays a key role in two other European projects, namely C-ITS Corridor and InterCor–Interoperable Corridors [82]. Within the InterCor framework (in which the other Member States involved are Belgium, France, and the UK), two relevant demonstration events held in the Netherlands should be mentioned. The first was a testfest, which took place in July 2017 and involved RWW, VSGN, and PVD services. It was followed by the GLOSA Pre-Testfest in June 2018, aimed at evaluating the interoperability level reached in implementing GLOSA services.

4.1.4. Belgium

The Belgium case presents two different contexts to be analysed: Flanders and Wallonia. The Flemish pilot site involves all the motorways that form part of the core network in Flanders. The area concerned is the Belgium site falling in the InterCor deployment framework. In the Flemish pilot, only Day-1 services are investigated, while the communication technology analysed concerns the implementation of cellular networks that are tested in combination with the HERE Location Cloud and the local Traffic Management Centre (TMC). The aim is to develop a cloud-based 'virtual infrastructure' connecting road users with the TMC while allowing the TMC to directly interact with end users. The implementing bodies are: the Flemish Department of Mobility and Public Works, Tractebel Engineering SA, ITS Belgium, and HERE Technologies. In accordance with the declared timeline, all preparatory phases (i.e., use case definition, system setup, fine-tuning, and driver acquisition) have been completed. The next phases consist in carrying out preliminary test drives and making the whole system operational, which will perform trial activities for the whole of 2019. Finally, the ex-post evaluation will be carried out in 2020.

As regards Wallonia test activities, the services and communication technologies tested as well as the planned timeline are essentially the same as the Flemish case, except for the fact that ITS-G5 is also investigated in Wallonia. The partners involved are: SOFICO (technically assisted by the Public Service of Wallonia/Directorate of Road Traffic Management), Tractebel Engineering SA, and ITS Belgium.

4.1.5. France

C-Roads France builds on the results of the European project SCOOP@F [83]. The goal is to test cooperative services within two types of end-user services: services in the urban environment and at the urban/interurban interface as well as traffic information services increasing comfort on transit stretches. Among Day-1 services, the majority of functions are investigated with the exception of some signage applications, namely VSPD, SWD, SigV, and TSP. Moreover, the development of a C-ITS smartphone application is planned, supporting early I2V (Infrastructure-to-Vehicle) services such as logistic services as well as parking and park and ride information. As regards the communication technology implemented, it is represented by a hybrid framework, enabling a seamless switch between ITS-G5 and cellular for non-safety-critical applications.

The partners involved are public road operators (DIRs EST, Centre-Est, Atlantique, and Ouest), road operator concessionaries (Autoroutes Paris-Rhin-Rhône–APRR, Société des Autoroutes du Nord et de l'Est de la France–SANEF, and VINCI Autoroutes), car manufacturers (Renault, PSA group), research institutes (CEREMA, IFSTTAR, Telecom Paris Tech, Université d'Auvergne Clermont-Ferrand, and Université de Reims Champagne-Ardennes), security experts (IDnomic), mobility labs (Car2road, Transpolis), and the major urban nodes authorities (Strasbourg Eurométropole, Bordeaux Métropole). According to the timeline stated, the rolling-out phase has just started. Finally, it is worth pointing out that France, in turn, is one of the Member States involved in the European project InterCor.

4.1.6. United Kingdom

C-Roads UK pilot builds on the core elements of the A2/M2 CVC–A2/M2 Connected Vehicle Corridor project [84], which, in turn, represents the UK part of the InterCor programme. The beneficiary is the UK Department for Transport (DfT) and the partners involved are Highways England (HE), Transport for London (TfL), and Kent County Council (KCC). In particular, HE and KCC intend to deliver their services through both ITS-G5 and cellular networks in a hybrid framework, whilst Transport for London intends to deliver the services via cellular technologies alone. According to the planned timeline, the four services GLOSA, IVS, RWW, and PVD are currently operational, and the demonstration will be held until May 2019. It is worth mentioning the demonstration event Hybrid Testfest, held in October 2018 in the UK, with the collaboration of other InterCor partners that aims at testing the implementation of the above Day-1 services within a hybrid G5/cellular framework.

4.1.7. Nordic Countries (Denmark, Sweden, Norway, and Finland)

Nordic countries are described in a unique, comprehensive framework since they are part of a joint initiative, namely the NordicWay project [85], and C-Roads pilot sites are essentially physically overlapping with those involved in the NordicWay programme. Strictly speaking, other countries are also involved in joint initiatives outlined in formally recognised European projects; however, in such cases, the European project and C-Roads tests differ, thus justifying the choice of describing the countries involved separately.

In light of the above, Denmark, Sweden, Norway, and Finland have a twofold aim: i) connecting their traffic management centres and their backbone road system with the common NordicWay architecture and ii) testing C-ITS applications in use-cases that are particularly relevant to their own national context.

In Denmark, C-Roads test activities, run by the Danish Road Directorate, are fully included in the NordicWay programme [86].

Also for Sweden, the limit between the NordicWay programme and C-Roads pilot sites is very vague; however, in this case, some Day-1.5 services (i.e., traffic information and smart routing; connected and cooperative navigation) are also being analysed. Finally, a large number of public and private organisations are involved in test activity under the leadership of the Swedish Transport Administration. Amongst others, we find: Telefonaktiebolaget LM Ericsson, Volvo Car Corporation, Chalmers University of Technology, KTH–Integrated Transport Research Lab (ITRL), Combitech AB, Triona AB, RISE Interactive Institute AB, Kapsch TrafficCom AB, Mindconnect AB, IBM Svenska AB, Swarco Sverige AB, Technolution AB, and Springworks AB.

In Norwegian pilot sites, two main goals are pursued under the leadership of Norwegian Public Roads Administration. The first is to test Day-1 and Day-1.5 services on peripheral networks, where rural routes have only poor cellular connectivity and no full access to main power. Moreover, connected and automated driving tasks on major freight routes are investigated. Finally, test activities within the Finnish pilot sites aim to investigate Day-1 and Day-1.5 services on the core network but also test automated driving in snowy and icy arctic conditions. In this case, under the leadership of the public authorities of the Finnish Transport Agency (FTA) and Finnish Transport Safety Agency (Trafi), studies are performed by three coalitions led by the Lapland University of Applied Sciences, Sensible 4 Ltd., and VTT Technical Research Centre of Finland.

Obviously, most of the deployment activities aim to achieve a high level of interoperability, among the Nordic countries themselves and with other European countries, by means of numerous cross-border tests. Indeed, as regards the communication technologies analysed, they are essentially cellular-based; however, for those services that can also be implemented with ITS-G5 standards, both technologies are tested in an interoperability perspective. Overall, the Nordic countries have covered all Day-1 applications, except for SWD services. Moreover, the Nordic case represents, among those presented, the first instance in which advanced C-ITS services (i.e., CCRW and CCN) are involved in testing activities. A specific description of C-ITS tasks investigated in every single country will be provided below. Finally, the above-mentioned pilot sites are planned to be operational for the whole of 2019.

4.1.8. Portugal

C-Roads Portugal involves the Atlantic Corridor in Portugal and covers major sections of the core and comprehensive networks, together with Lisbon and Porto urban nodes. In particular, the programme is developed through five pilot sites whose activities are specified below.

- Pilot 1 for designing a National Single Point of Access (SPA) prototype able to cover information for around 3390 km (20%) of the network and developing an SPA mobile application (SPApp) covering Day-1 services
- Pilot 2 for testing Day-1 and Day-1.5 on different kinds of roads (metropolitan areas, interurban roads, streets, and highways) using a hybrid G5/cellular communication system. Pilot activities cover over 460 km of the core and comprehensive network, including cross-border sections in Valença and Caia and roads giving access to urban nodes of Lisbon and Porto
- Pilot 3 for providing connected and autonomous vehicles with automation on levels two and three of the Trans-European Networks-Transport (TEN-T) network, also using a hybrid G5/cellular communication framework
- Pilot 4 consisting of the following subactivities focused on the Lisbon node: testing traffic monitoring and travel time prediction tasks by means of cellular technology; investigating infotainment services on parking availability with, in addition, the development of an in-vehicle app based on a hybrid communication framework; analysing bus corridor prioritisation services supported by cellular technologies; and evaluating potential benefits of the integration of private car usage with other transport modes in the last mile of interurban motorway corridors in a hybrid communication architecture
- Pilot 5 for investigating applications on traffic prediction services, by means of cellular and Wi-fi technologies, and testing the incorporation in the Porto network of a smart bus covering a corridor of around 1.4 km, using the DATEXII communications protocol and cellular communication technologies

More than 30 partners are involved, including: Instituto da Mobilidade e dos Transportes, Universidade do Porto, Ascendi Grande Lisboa, Ascendi do Grande Porto, IP Telecom SA, Siemens SA, Vialivre SA, and GMVIS Skysoft SA. According to the planned timeline, by the end of 2020, C-ITS services will be implemented along 964 km, ensuring a continuity of service in urban nodes and the core network. In all, 212 RSUs will be installed with 180 OBUs and 162 vehicles in operation.

4.1.9. Spain

Similar to Portugal, C-Roads Spain develops through five pilot sites:

- DGT 3.0—located along the overall road network in Spain, with an extension of approximately 12,270 km. It will be deployed by adopting cellular-based communication technologies (3G and 4G/LTE).
- SISCOGA Extended—including the extension of an existing test site infrastructure in the city of Vigo and its metropolitan area, which is already prepared to test ITS-G5 communication technology. It will cover 150 km.
- Madrid Calle 30—located along the road 'Calle 30' in Madrid, approximately 32 km long. C-ITS services will be deployed by using a hybrid G5/cellular approach.
- Cantabrian pilot—deployed along approximately 75 km in northern Spain, by using hybrid communications.
- Mediterranean pilot—deployed along approximately 125 km at selected road sections in Catalonia and Andalucia using hybrid technologies

Such areas present a high level of heterogeneity, thus allowing a wide spectrum of use-cases to be tested involving almost all Day-1 applications, except for TJW services, and various Day-1.5 functions, namely VRU, CCN, and infotainment tasks. Moreover, cross-border tests with Portugal are also planned within the SISCOGA Extended pilot site.

A large number of partners are involved, including public authorities (Dirección General de Tráfico, Dirección General de Carreteras, and Madrid Calle 30 SA), associations (ITS Spain, MLC ITS EUSKADI), private companies (Transport Simulation Systems SL, Indra Systemas SA, Ingartek Consulting SL, and SenseFields SL), Universities (Universidad Politécnica de Madrid, Universitat Politècnica de Catalunya, and Universitat de Valencia) and research centres (Asociación Centro Tecnológico Ceit-IK4, CTAG). According to the planned timeline, the programme will be rolled out during 2019 in two phases, while 2020 will be devoted to evaluation.

4.1.10. Czech Republic

Czech Republic activities, within the C-Roads programme, are outlined in six different pilot sites, from deployment and test (DT)1 to DT6. The services tested cover almost all Day-1 applications, excluding only shockwave damping and GLOSA tasks, and focus particularly on public transport safety. Both ITS-G5 and cellular-based technologies are being investigated with the aim of developing a hybrid framework. Several implementing bodies among transport operators, research organisations, and equipment supplier companies are involved, namely Road and Motorway Directorate (RSD), Brněnské komunikace company, Správa Železniční Dopravní Cesty (SŽDC) railway operator, public transport companies of Ostrava and Plzeň, INTENS Corporation, AŽD Praha, O2, T-Mobile, Škoda Auto, and Czech Technical University.

According to schedule, pilot sites from DT1 to DT5 were completed in 2018, while DT6 has just been activated. Therefore, all DT will be operational throughout 2019, and 2020 will be dedicated to ex-post evaluation.

4.1.11. Hungary

C-Roads Hungary project is focused on testing Day-1 services, while a hybrid DSRC/cellular technology is planned to be developed in the near future. The beneficiary is the Ministry of National Development (MND), while the implementing body is the Hungarian Public Road Nonprofit PLC. Other partners planned to be involved are: Budapest University of Technology and Economics (BUTE), Budapest Public Road PLC, Automotive Proving Ground Zala LTD (APZ), and ITS Hungary Association. According to the declared timeline, the test phase started in the second half of 2018 and is planned to last until the end of 2020, with an upgrade of equipment scheduled by the end of 2019. Finally, cross-border tests are also envisaged in 2020.

4.1.12. Italy

C-ROADS Italy project is focused on the following use-cases: anticipated disengagement of vehicle L3 highway pilot, truck platooning, passenger cars highway chauffeur, and combined scenarios of trucks and passenger cars.

The coordinator is the Italian Ministry of Infrastructure and Transport, which, at the beginning of 2018, set up a monitoring body exclusively dedicated to cooperative driving tasks. Among the implementing bodies there are different organisations, including: motorway operators (i.e., Brenner Motorway SPA, Autovie Venete Motorway SPA, and CAV–Concessioni Autostradali Venete Motorway SPA), carmakers (Iveco, PSA Group, Renault, Volvo, and Scania), research bodies (the Polytechnic University of Milan and FCA Research Centre), communication and logistic industries (North Italy Communications SRL, Telecom Italia SPA, Azcom Technology SRL, and Codognotto SNC). According to the planned timeline, by the end of 2019, 50 km of the Brenner Motorway, 10 km of the Autovie Venete Motorway, and 10 km of the CAV Motorway will be operational. Moreover, cross-border tests with Austria are planned for 2020. Specifically, ITS-G5 and cellular communication technologies will be tested.

4.1.13. Slovenia

C-Roads Slovenia pilot sites have been developed in two phases. The first, in which both ITS-G5 and cellular networks are tested, is planned along 100 km of the TEN-T core network. In the second phase, planned to start in 2020, roadside C-ITS-G5 infrastructure will be extended to the pilot length of 300 km, including critical points connected with the Central C-ITS-G5 Server real-time platform located at the Dragomelj Traffic Management Centre. The implementing bodies are the Ministry of Infrastructure and the Motorway Company of the Republic of Slovenia (DARS). The tested services are all Day-1 type, namely SSV, TJM, RWW, WTC, VSGN, VSPD, and GLOSA applications.

4.1.14. Summary

A synopsis of C-ITS services tested in each Member State within the C-Roads programme is shown in Table 2, giving a state-of-the-art overview of the Day-1 and Day-1.5 applications investigated.

Table 2. C-ITS services tested in C-Roads pilot sites.

		C-ITS Services																		
		Day-1										Day-1.5								
		EBL	EVA	SSV	TJW	RWW	WTC	VSGN	VSPD	PVD	SWD	GLOSA	Sig	TSP	Info	LZMC	ZAR	VCU	CCRW	MAD	WCD	CCN
C-ITS Corridor	Austria			✓	✓	✓	✓	✓	✓	✓												
	Germany	✓	✓	✓	✓		✓			✓	✓	✓										
	The Netherlands				✓		✓			✓		✓			✓							
InterCor	Belgium			✓	✓	✓	✓		✓		✓											
	France	✓	✓	✓	✓	✓	✓	✓		✓		✓			✓							
	United Kingdom				✓		✓			✓		✓										
NordicWay	Denmark			✓	✓	✓	✓			✓												
	Finland	✓	✓		✓	✓	✓	✓	✓		✓	✓	✓		✓							
	Norway	✓		✓		✓	✓	✓	✓		✓	✓			✓		✓					
	Sweden	✓	✓		✓		✓					✓		✓								✓
	Portugal	✓	✓	✓	✓	✓	✓	✓	✓	✓	✓	✓	✓		✓	✓				✓	✓	
	Spain	✓	✓	✓		✓	✓	✓	✓	✓	✓	✓	✓		✓		✓					✓
	Czech Republic	✓	✓	✓	✓	✓	✓	✓	✓	✓			✓	✓								
	Hungary				✓	✓	✓	✓	✓	✓			✓	✓								
	Italy	✓			✓	✓	✓	✓	✓	✓	✓											
	Slovenia				✓	✓	✓	✓	✓	✓			✓									

In particular, as can be seen, Day-1 services are implemented in the majority of EU countries (16 over 28). Moreover, it is worth noting that, for VSGN, GLOSA, RWW, and other hazardous location notification applications, shared deployment principles and specifications are established for assuring interoperability across the Union [87]. Table 2 also shows that Day-1.5 services are less widely investigated since they are not undergoing tests at all in 9 out of 16 Member States, as highlighted by grey lines in the table; if infotainment tasks are excluded, the proportion drops to a quarter. Information in Table 2 could also be combined with those in Figure 2, which provides a synoptic view about where C-ITS service deployment activities are located across Europe and the magnitude of tests performed. In particular, black regions and lines have been adopted for indicating areas in which a greater number of C-ITS services is evaluated. Clearly, it represents only a drafted illustration, since the goal is to provide an at-a-glance view of the current situation. Moreover, it is worth specifying that only activities having an importance from a pan-European point of view have been considered, while initiatives of national importance (e.g., the Smart Roads project in Italy or the Human Drive initiative in UK) have been neglected. However, with all the previous in mind, it can be seen that the most intense test activity occurs on TEN-T corridors that, not surprisingly, have been included in formally recognised European projects such as C-ITS corridor and InterCor. In particular, the involvement in deployment activities (and possibly in the level of funding distribution) is very heterogeneous across different countries in the EC. Southern Europe and some parts of Eastern Europe are completely excluded, and some differences are very undeniable among involved states. For example, Portugal and Italy are in a totally different situation, with the former having all the most relevant infrastructures under development and the latter being involved in testing activities only for a small part of the country. Furthermore, within the NordicWay project, Sweden and Denmark are less proactive than other partners, and the latter is the only partner country in which exclusively Day-1 services are under test.

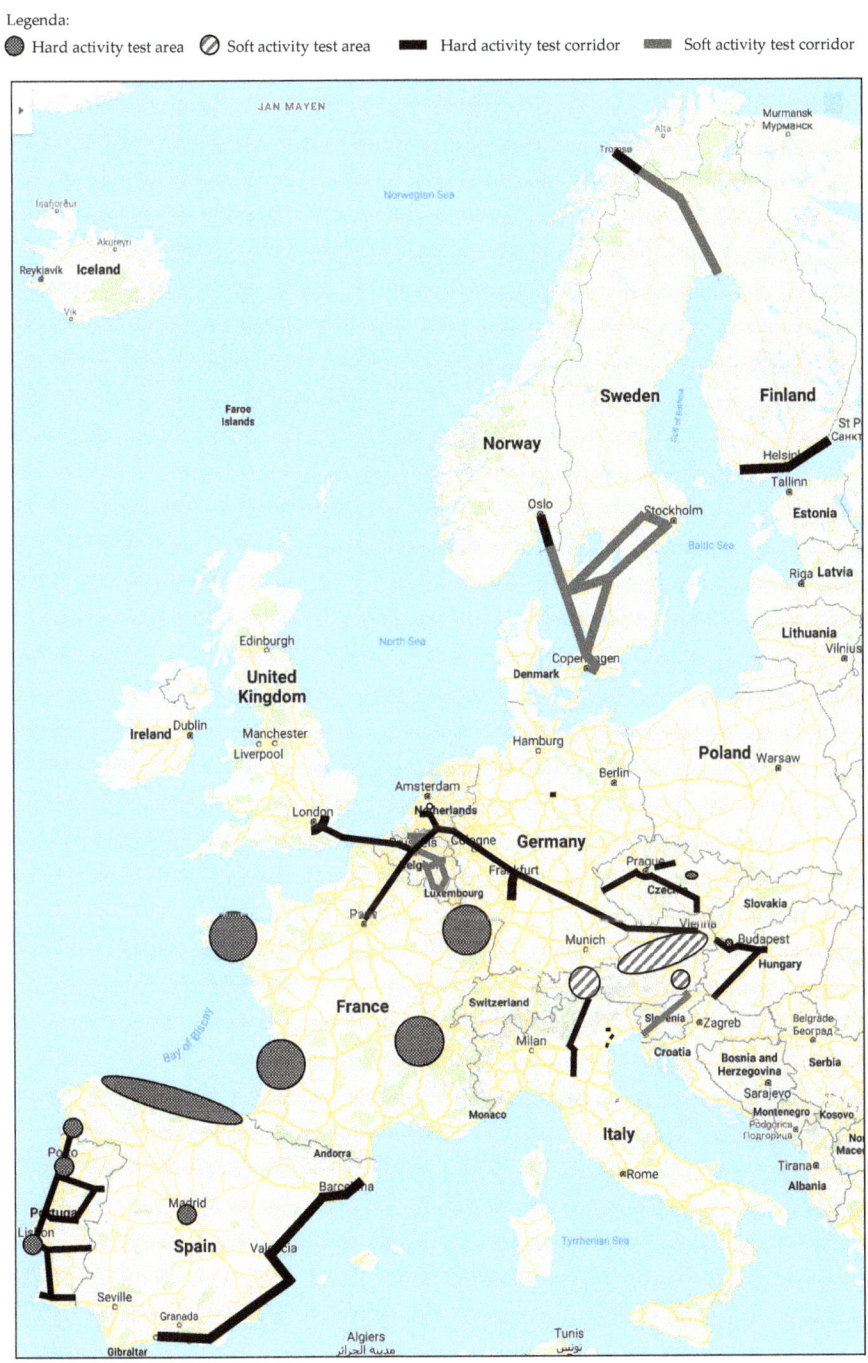

Figure 2. Synoptic view of C-ITS deployment activities across the EU.

Regarding communication technologies, it is worth specifying that although ITS-G5 and cellular-based communications represent the most mature technologies, additional systems are being tested such as Wi-Fi and Bluetooth (in France, the Czech Republic, and Portugal), radio data (in Portugal), and digital audio broadcasting (in Germany).

Finally, analysing the types of stakeholders involved in the C-Roads programme [88], a considerable heterogeneity is observable across Member States. The number of entities involved is obviously consistent with the quantity of actions carried out, ranging from 1–2 (in the case of Austria and Slovenia) to over 30 (for Portugal and Sweden). The actual drivers of the development process in different countries are also quite heterogeneous. In fact, countries with a well-established tradition in the automotive sector (such as Germany, Sweden, and France) present considerable involvement of car-makers and equipment suppliers gravitating around the automobile industry. By contrast, countries like Austria, the Netherlands, the UK, and Denmark show a more centralised management behaviour, with public authorities being the main players. However, as can be seen, the most diverse organisations are involved, thus confirming the need to adopt a multidisciplinary approach for successful deployment of C-ITS applications.

5. Concluding Remarks

The above overview provides an insight into the current deployment activities in the European Union within the field of C-ITS and V2X communication. Best practices across the European Union were investigated, with the aim of outlining the state of play in early 2019, which was identified as the start time for deployment of mature services.

In sum, against a total budget of around EUR 300 million allocated for cooperative driving development by the EU's framework programme Horizon 2020, a resulting picture emerges with the following highlights. Regarding infrastructural pilot sites, a common trend can be identified. Indeed, generally from 2016 to 2018, preparatory phases were performed such as use-case definition, specification, development, and validation. Thus, they will be ready to start rolling out in 2019 at the latest and will be devoted to evaluation in 2020. The most mature aspect is the technological one, where ITS-G5 and cellular-based communications have been shown to be ready for implementation, possibly in different contexts and use-cases. Hybrid frameworks have been tested as well, allowing implementation of C-ITS services to potentially benefit from the advantages offered by both communication systems. By contrast, with reference to C-ITS services viewed from an applicative perspective involving final users/drivers, a fairly heterogeneous development can be noted; RWW appears as the only service investigated in almost all the testbeds, followed by PVD, SSV, and VSGN. We identified three possible reasons for such a different state of play among the above-mentioned C-ITS applications.

First, vehicle-side technological and applicative issues could delay the deployment of some tasks. In this regard, it may be stated that the use of cameras, radar transmitters, sensors, laser scanners, and digital maps appears consolidated. Moreover, several ADASs, such as adaptive cruise control, emergency braking, backup cameras, and self-parking systems, have already been installed on commercial vehicles. However, in order to take a step forward and enable more complex tasks, as shown by [89], systematisation and harmonisation of issues related to perception, localisation, and the decision-making process in a fail-safe scenario are required. Perception issues concern the development of suitable algorithms able to read the surrounding environment with a high degree of reliability in all operational domains, distinguishing, for instance, a stationary motorcycle from a bicyclist riding at the roadside. Essentially, object analysis is made complex by the various randomnesses involved (e.g., movement, background, and weather conditions), which account for the failure of most test activities and at times also lead to fatal crashes (even not in C-ITS testbeds). As regards the question of localisation, it is related to the error range of the GPS signal. In this regard, the EU undertook further development of Galileo services and their accuracy for driverless mobility. Finally, the human decision-making process should be suitably modelled in a software environment, thus reproducing driver choice behaviour. This is undoubtedly the most complex aspect to be addressed.

The latter issue leads us to the second possible reason for the lack of development in certain C-ITS services, which is a methodological issue (i.e., the lack of a robust procedure for successfully finalising the task). This happens, for instance, in the case of advanced driving information tasks. Indeed, although the technology for acquiring and delivering information is mature, the extremely dynamic nature of such a task means that services concerning, for instance, traffic information and smart routing, even if increasingly popular, are not yet consolidated from the point of view of their ability to be predictive and pre-emptive. Indeed, after receiving certain information with traffic forecasting characteristics, users change their behaviour, thus altering the utility and reliability of the same information for the rest of the drivers [90]. There is no such complication, instead, when stationary objects are involved, such as the case of road works. Therefore, as shown by [91], the following research questions are still being debated: How do drivers change their behaviour because of warnings/information given by C-ITS services? Is safety affected by changes in driver behaviour because of C-ITS services?

The third reason is related to the spread of equipped C-ITS systems. Indeed, issues concerning the standardisation of information and procedures can be identified in the case of services involving a large number of vehicles or different traffic participants such as shockwave damping, vulnerable road user protection, and connected and cooperative navigation.

Our overview, albeit comprehensive, was not exhaustive: it mainly focused on roadside deployment activities. However, in the field of V2X communication, infrastructure equipment and implementation of in-vehicle technologies represent two sides of the same coin. They are complementary to one another, as demonstrated also by the C2C roadmap framework, and need to be deployed synergistically. Nevertheless, if roadside deployment can be handled in a strategic perspective, with standardisation and regulation principles dictated from a centralised authority, the carmakers side will present a more fragmentary framework affected by particular aspects of each company, such as the know-how possessed and in-house policies adopted. Another limitation of the presented work concerns the initiatives mentioned and related stakeholders. Indeed, given the extent of the phenomenon as well as its ever-changing nature, it is next to impossible to provide a complete list of the countless deployment activities and public/private organisations involved. Therefore, far from any claim to be exhaustive, our contribution aims to provide a valid synopsis of the current state of affairs so as to steer future efforts along the right lines.

Finally, as shown by the study described in [92], when autonomous vehicles become available, a sizable increase occurs in vehicle miles travelled and number of trips. This is hardly surprising: it would be as if each user had a personal chauffeur. Such an event could negatively affect benefits in terms of sustainability and congestion reduction, of which mention is generally made. Hence, the authors suggest including in the evaluation step a study focusing on the above point, thus providing a realistic and reliable representation of what the future with autonomous vehicles would look like.

Author Contributions: Conceptualization, M.B., L.P. and G.N.B.; Methodology, L.D. and G.N.B., Investigation, M.B.; Data curation, M.B. and L.P.; Visualization, M.B. and L.P.; Writing–original draft preparation, M.B., L.P., L.D. and G.N.B.; Writing–review and editing, M.B., L.P., L.D. and G.N.B.; Supervision, L.P. L.D. and G.N.B.; Validation, L.D. and G.N.B.

Funding: This research was partially funded by Federico II University of Naples (Italy), grant number 000009-ALTRO_R-2016-LD "*Models and technologies for transportation systems*".

Conflicts of Interest: The authors declare no conflict of interest. The funders had no role in the design of the study; in the collection, analyses, or interpretation of data; in the writing of the manuscript, or in the decision to publish the results.

Abbreviations

A2/M2 CVC	A2/M2 Connected Vehicle Corridor
ADAS	Advanced Driver Assistance Systems
AG	Amsterdam Group
APZ	Automotive Proving ground Zala
BASt	Federal Highway Research Institute
BUTE	Budapest University of Technology and Economics
C2C-CC	Car2Car Communication Consortium
CACC	Cooperative Adaptive Cruise Control
CAM	Cooperative Awareness Message
CAV	Concessioni Autostradali Venete
CCAM	Cooperative Connected and Automated Mobility
CCN	Connected and Cooperative Navigation
CCRW	Cooperative Collision Risk Warning
CELC	Cooperative Emergency Lane Change
C-ITS	Cooperative-Intelligent Transportation Systems
CTAG	Centro Tecnológico de Automoción de Galicia
DENM	Decentralized Environmental Notification Message
DfT	Department for Transport
DSRC	Dedicated Short-Range Communication
DT	Deployment and Tests
EBL	Emergency electronic Brake Light
EC	European Commission
ECo-AT	European Corridor Austrian Testbed
ETSI	European Committee for Standardisation
EU	European Union
EVA	Emergency Vehicle Approaching
FMI	Finnish Meteorological Institute
FTA	Finnish Transport Agency
GLOSA	Green Light Optimal Speed Advisory
GUI	Graphical User Interface
HE	Highways England
HGV	Heavy Goods Vehicle
HLG	High Level Group
HLN	Hazardous Location Notifications
HMI	Human Machine Interface
I2V	Infrastructure-to-Vehicle
ITRL	Integrated Transport Research Lab
IVI	In-Vehicle Information
KCC	Kent County Council
LZM	Loading Zone Management
MAC	Medium Access Control
MAP	Map Data
MCA	Motorcycle Approaching indication
MCTO	Multimodal Cargo Transport Optimization
MND	Ministry of National Development
MPC	Model Predictive Controller
NR	New Radio
OBU	On-Board Unit
OEMs	Original Equipment Manufacturers
PACE	Parking Autonomously in Cooperative Environments
PHY	Physical
PVD	Probe Vehicle Data
RSD	Road and Motorway Directorate
RSU	Road-Side Unit

RWS	Road Weather Station
RWW	Road Works Warning
SAE	Society of Automotive Engineers
SCMA	Sparse Coded Multiple Access
SDR	Software Defined Radio
SigV	Signal Violation/ Intersection safety
SPA	Single Point of Access
SPApp	Single Point of Access Application
SPaT	Signal Phase and Timing
SSV	Slow or Stationary Vehicle
STLM	Smart Traffic Light Manager
SWD	Shockwave Damping
SŽDC	Správa Železniční Dopravní Cesty
TEN-T	Trans-European Networks - Transport
TfL	Transport for London
TJW	Traffic Jam ahead Warning
TLPM	Traffic Light Power Manager
TMC	Traffic Management Centre
TPMT	Test site Project Management Team
Trafi	Finnish Transport Safety Agency
TSP	Traffic Signal Priority
TTG	Time to Green
URLLC	Ultra-Reliable and Low-Latency Communication
V2I	Vehicle-to-Infrastructure
V2P	Vehicle-to-Pedestrian
V2V	Vehicle-to-Vehicle
V2X	Vehicle-to-Everything
VNF	Virtualized Network Function
VRU	Vulnerable Road User
VSGN	in-Vehicle Signage
VSPD	in-Vehicle Speed limits
WTC	Weather Conditions
WWD	Wrong Way Driving
ZAC	Zone Access Control

References

1. *Commission Delegated Regulation of 13.3.2019 Supplementing Directive 2010/40/EU of the European Parliament and of the Council with Regard to the Deployment and Operational Use of Cooperative Intelligent Transport Systems;* European Commission: Brussels, Belgium, 2019; Available online: http://ec.europa.eu/transparency/regdoc/rep/3/2019/EN/C-2019-1789-F1-EN-MAIN-PART-1.PDF (accessed on 30 April 2019).
2. European Commission, Cooperative, Connected and Automated Mobility (CCAM). Available online: ec.europa.eu/transport/themes/its/c-its_en (accessed on 30 April 2019).
3. Car2Car Communication Consortium (C2C-CC). Available online: www.car-2-car.org (accessed on 30 April 2019).
4. Amsterdam Group (AG). Available online: amsterdamgroup.mett.nl (accessed on 30 April 2019).
5. Botte, M.; Pariota, L.; D'Acierno, L.; Bifulco, G.N. C-ITS communication: An insight on the current research activities in the European Union. *J. Transp. Syst.* **2018**, *3*, 52–63.
6. Society of Automotive Engineers (SAE). *Taxonomy and Definitions for Terms Related to Driving Automation Systems for On-Road Motor Vehicles.* Standard SAE J3016-2018. 2018. Available online: https://www.sae.org/standards/content/j3016_201806 (accessed on 30 April 2019).
7. *On the Road to Automated Mobility: An EU Strategy for Mobility of the Future;* European Commission: Brussels, Belgium, 2018; Available online: https://ec.europa.eu/transport/sites/transport/files/3rd-mobility-pack/com20180283_en.pdf (accessed on 30 April 2019).

8. *Ensuring That Europe Has The Most Competitive, Innovative and Sustainable Automotive Industry of the 2030s and Beyond.* GEAR 2030 Final Report. 2017. Available online: https://ec.europa.eu/docsroom/documents/26081/attachments/1/translations/en/renditions/pdf (accessed on 30 April 2019).
9. European Telecommunications Standards Institute (ETSI). *Intelligent Transport Systems (ITS); Vehicular Communications; Basic Set of Applications; Definitions.* Standard ETSI TR 102 638-1. 2009. Available online: www.etsi.org/deliver/etsi_tr/102600_102699/102638/01.01.01_60/tr_102638v010101p.pdf (accessed on 30 April 2019).
10. European Telecommunications Standards Institute (ETSI). *Intelligent Transport Systems (ITS); Infrastructure to Vehicle Communication; Electric Vehicle Charging Spot Notification Specification.* Standard ETSI TS 101 556-1. 2012. Available online: https://www.etsi.org/deliver/etsi_ts/101500_101599/10155601/01.01.01_60/ts_10155601v010101p.pdf (accessed on 30 April 2019).
11. International Organization for Standardization (ISO). *Intelligent Transport Systems; Event-Based Probe Vehicle Data.* Standard ISO/TS 29284. 2012. Available online: www.iso.org/obp/ui#iso:std:iso:ts:29284:ed-1:v1:en (accessed on 30 April 2019).
12. European Telecommunications Standards Institute (ETSI). *Intelligent Transport Systems (ITS); V2X Applications; Part 1: Road Hazard Signalling (RHS) Application Requirements Specification.* Standard ETSI TS 101 539-1. 2013. Available online: www.etsi.org/deliver/etsi_ts/101500_101599/10153901/01.01.01_60/ts_10153901v010101p.pdf (accessed on 30 April 2019).
13. International Organization for Standardization (ISO). *Intelligent Transport Systems; Forward Vehicle Collision Warning Systems; Performance Requirements and Test Procedures.* Standard ISO 15623. 2013. Available online: www.iso.org/obp/ui/#iso:std:iso:15623:ed-2:v1:en (accessed on 30 April 2019).
14. International Organization for Standardization (ISO). *Intelligent Transport Systems (ITS); Cooperative Intersection Signal Information and Violation Warning Systems (CIWS); Performance Requirements and Test Procedures.* Standard ISO 26684. 2015. Available online: www.iso.org/obp/ui#iso:std:iso:26684:ed-1:v1:en (accessed on 30 April 2019).
15. International Organization for Standardization (ISO). *Intelligent Transport Systems; Cooperative Systems; Data Exchange Specification for In-Vehicle Presentation of External Road and Traffic Related Data.* Standard ISO/TS 17425. 2016. Available online: www.iso.org/obp/ui/#iso:std:iso:ts:17425:ed-1:v1:en (accessed on 30 April 2019).
16. European Telecommunications Standards Institute (ETSI). *Intelligent Transport Systems (ITS); V2X Applications; Part 2: Intersection Collision Risk Warning (ICRW) Application Requirements Specification.* Standard ETSI TS 101 539-2. 2018. Available online: www.etsi.org/deliver/etsi_ts/101500_101599/10153902/01.01.01_60/ts_10153902v010101p.pdf (accessed on 30 April 2019).
17. European Telecommunications Standards Institute (ETSI). *Intelligent Transport Systems (ITS); Vehicular Communications; Basic Set of Applications; Part 2: Specification of Cooperative Awareness Basic Service.* Standard ETSI EN 302 637-2. 2014. Available online: www.etsi.org/deliver/etsi_en/302600_302699/30263702/01.03.01_30/en_30263702v010301v.pdf (accessed on 30 April 2019).
18. European Telecommunications Standards Institute (ETSI). *Intelligent Transport Systems (ITS); Vehicular Communications; Basic Set of Applications; Part 3: Specifications of Decentralized Environmental Notification Basic Service.* Standard ETSI EN 302 637-3. 2014. Available online: www.etsi.org/deliver/etsi_en/302600_302699/30263703/01.02.01_30/en_30263703v010201v.pdf (accessed on 30 April 2019).
19. Santa, J.; Pereñíguez, F.; Moragón, A.; Skarmeta, A.F. Vehicle-to-Infrastructure messaging proposal based on CAM/DENM specifications. In Proceedings of the 6th IFIP/IEEE Wireless Days Conference—WDays 2013, Valencia, Spain, 13–15 November 2013. [CrossRef]
20. European Telecommunications Standards Institute (ETSI). *Intelligent Transport Systems (ITS); Testing; Conformance Test Specifications for Signal Phase and Timing (SPAT) and Map (MAP) Part 1: Test Requirements and Protocol Implementation Conformance Statement (PICS) pro Forma.* Standard ETSI TS 103 191-1. 2015. Available online: www.etsi.org/deliver/etsi_ts/103100_103199/10319101/01.01.01_60/ts_10319101v010101p.pdf (accessed on 30 April 2019).
21. European Telecommunications Standards Institute (ETSI). *Intelligent Transport Systems (ITS); Testing; Conformance Test Specifications for Signal Phase And Timing (SPAT) and Map (MAP) Part 2: Test Suite Structure and Test Purposes (TSS&TP).* Standard ETSI TS 103 191-2. 2015. Available online: www.etsi.org/deliver/etsi_ts/103100_103199/10319102/01.01.01_60/ts_10319102v010101p.pdf (accessed on 30 April 2019).

22. European Telecommunications Standards Institute (ETSI). *Intelligent Transport Systems (ITS); Testing; Conformance test Specifications for Signal Phase and Timing (SPAT) and Map (MAP) Part 3: Abstract Test Suite (ATS) and Protocol Implementation eXtra Information for Testing (PIXIT)*. Standard ETSI TS 103 191-3. 2015. Available online: www.etsi.org/deliver/etsi_ts/103100_103199/10319103/01.01.01_60/ts_10319103v010101p.pdf (accessed on 30 April 2019).
23. International Organization for Standardization (ISO). *Intelligent Transport Systems; Cooperative ITS; Dictionary of In-Vehicle Information (IVI) Data Structures*. Standard ISO/TS 19321. 2015. Available online: www.iso.org/obp/ui/#!iso:std:iso:ts:19321:ed-1:v1:en (accessed on 30 April 2019).
24. European Telecommunications Standards Institute (ETSI). *Intelligent Transport Systems (ITS); Vehicular Communications; Basic Set of Applications; Facilities Layer Protocols and Communication Requirements for Infrastructure Services*. Standard ETSI TS 103 301. 2016. Available online: https://www.etsi.org/deliver/etsi_ts/103300_103399/103301/01.01.01_60/ts_103301v010101p.pdf (accessed on 30 April 2019).
25. European Telecommunications Standards Institute (ETSI). *Intelligent Transport Systems (ITS); European Profile Standard for the Physical and Medium Access Control Layer of Intelligent Transport Systems Operating in the 5 GHz Frequency Band*. Standard ETSI ES 202 663. 2012. Available online: www.etsi.org/deliver/etsi_es/202600_202699/202663/01.01.00_50/es_202663v010100m.pdf (accessed on 30 April 2019).
26. Institute of Electrical and Electronics Engineers (IEEE). *Wireless LAN Medium Access Control (MAC) and Physical Layer (PHY) Specifications*. Standard IEEE 802.11. 2007. Available online: https://www.iith.ac.in/~{}tbr/teaching/docs/802.11-2007.pdf (accessed on 30 April 2019).
27. Institute of Electrical and Electronics Engineers (IEEE). *Wireless LAN Medium Access Control (MAC) and Physical Layer (PHY) Specifications—Amendment 6: Wireless Access in Vehicular Environments*. Standard IEEE 802.11p. 2010. Available online: www.ietf.org/mail-archive/web/its/current/pdfqf992dHy9x.pdf (accessed on 30 April 2019).
28. Sjöberg, K.; Andres, P.; Buburuzan, T.; Brakemeier, A. Cooperative Intelligent Transport Systems in Europe. *IEEE Veh. Technol. Mag.* **2017**, *12*, 89–97. [CrossRef]
29. Filippi, A.; Moerman, K.; Daalderop, G.; Alexander, P.D.; Schober, F.; Pfliegl, W. *Ready to Roll: Why 802.11p Beats LTE and 5G for V2X*. NXP Semiconductors, Cohda Wireless and Siemens White Paper. 2016. Available online: https://www.siemens.com/content/dam/webassetpool/mam/tag-siemens-com/smdb/mobility/road/connected-mobility-solutions/documents/its-g5-ready-to-roll-en.pdf (accessed on 30 April 2019).
30. European Telecommunications Standards Institute (ETSI). *LTE; Service Requirements for V2X Services (3GPP TS 22.185 Version 14.3.0 Release 14)*. Standard ETSI TS 122 185. 2017. Available online: www.etsi.org/deliver/etsi_ts/122100_122199/122185/14.03.00_60/ts_122185v140300p.pdf (accessed on 30 April 2019).
31. European Telecommunications Standards Institute (ETSI). *5G; Service Requirements for Enhanced V2X Scenarios (3GPP TS 22.186 Version 15.3.0 Release 15)*. Standard ETSI TS 122 186. 2018. Available online: https://www.etsi.org/deliver/etsi_ts/122100_122199/122186/15.03.00_60/ts_122186v150300p.pdf (accessed on 30 April 2019).
32. 5G Automotive Association. *The Case for Cellular V2X for Safety and Cooperative Driving*. White Paper. 2016. Available online: http://5gaa.org/wp-content/uploads/2017/10/5GAA-whitepaper-23-Nov-2016.pdf (accessed on 30 April 2019).
33. Flament, M. *Path Towards 5G for the Automotive Sector*. 5G Automotive Association Presentation. 2018. Available online: http://www.3gpp.org/ftp/Information/presentations/presentations_2018/2018_10_17_tokyo/presentations/2018_1017_3GPP%20Summit_07_5GAA_FLAMENT.pdf (accessed on 30 April 2019).
34. Rebbeck, T.; Stewart, J.; Lacour, H.A.; Killeen, A.; McClure, D.; Dunoyer, A. *The Cost-Benefit Analysis on Cellular Vehicle-to-Everything (C-V2X) Technology and Its Evolution to 5G-V2X*. 5G Automotive Association Report. 2017. Available online: http://5gaa.org/wp-content/uploads/2017/12/Final-report-for-5GAA-on-cellular-V2X-socio-economic-benefits-051217_FINAL.pdf (accessed on 30 April 2019).
35. 5G Americas. *Cellular V2X Communications Towards 5G*. White Paper. 2018. Available online: http://www.5gamericas.org/files/9615/2096/4441/2018_5G_Americas_White_Paper_Cellular_V2X_Communications_Towards_5G__Final_for_Distribution.pdf (accessed on 30 April 2019).
36. Fallgren, M.; Dillinger, M.; Alonso-Zarate, J.; Boban, M.; Abbas, T.; Manolakis, K.; Mahmoodi, T.; Svensson, T.; Laya, A.; Vilalta, R. Fifth-generation technologies for the connected car: Capable systems for Vehicle-to-Anything communications. *IEEE Veh. Technol. Mag.* **2018**, *13*, 28–38. [CrossRef]

37. Härri, J.; Brens, F. Challenges and Opportunities of WiFi-based V2X Communications. In Proceedings of the VDI Conference on Digital Infrastructure & Automotive Mobility, Berlin, Germany, 5–6 July 2017.
38. Papathanassiou, A.T.; Khoryaev, A. Cellular V2X as the Essential Enabler of Superior Global Connected Transportation Services. *IEEE 5G Tech Focus* **2017**, *1*. Available online: https://futurenetworks.ieee.org/tech-focus/june-2017/cellular-v2x (accessed on 30 April 2019).
39. Turley, A.; Moerman, K.; Filippi, A.; Martinez, V. *C-ITS: Three Observations on LTE-V2X and ETSI ITS-G5—A Comparison*. NXP Semiconductors White Paper. 2018. Available online: https://www.nxp.com/docs/en/white-paper/CITSCOMPWP.pdf (accessed on 30 April 2019).
40. *C-ITS Platform Phase II Cooperative Intelligent Transport Systems Towards Cooperative, Connected and Automated Mobility*. European Commission Report. 2017. Available online: https://ec.europa.eu/transport/sites/transport/files/2017-09-c-its-platform-final-report.pdf (accessed on 30 April 2019).
41. AnaVAnet Project. Available online: http://anavanet.net/ (accessed on 30 April 2019).
42. ESEMBLE Project. Available online: http://platooningensemble.eu (accessed on 30 April 2019).
43. MAVEN Project. Available online: http://www.maven-its.eu (accessed on 30 April 2019).
44. 5Gcar Project. Available online: https://5gcar.eu (accessed on 30 April 2019).
45. AUTOPILOT Project. Available online: http://autopilot-project.eu (accessed on 30 April 2019).
46. 5GINFIRE: IT-AV Automotive Environment. Available online: https://5ginfire.eu/it-av-automotive-testbed/ (accessed on 30 April 2019).
47. Telefonica and Huawei: 5G-V2X Testbed. EuropaWire, 2018. Available online: https://news.europawire.eu/telefonica-and-huawei-complete-joint-5g-v2x-poc-test-in-their-5g-joint-innovation-lab-at-madrid-53202031254/eu-press-release/2018/02/08/ (accessed on 30 April 2019).
48. Hecker, T.; Zech, J.; Schäufele, B.; Gräfe, R.; Radusch, I. Model car testbed for development of V2X applications. *J. Commun.* **2011**, *6*, 115–124. [CrossRef]
49. Cao, H.; Gangakhedkar, S.; Ramadam Ali, A.; Gharba, M.; Eichinger, J. A testbed for experimenting 5G-V2X requiring Ultra Reliability and Low-Latency. In Proceedings of the 21st International ITG Workshop on Smart Antennas—WSA 2017, Berlin, Germany, 15–17 March 2017.
50. Varga, N.; Bokor, L.; Takacs, A.; Kovacs, J.; Virag, L. An architecture proposal for V2X communication-centric traffic light controller systems. In Proceedings of the 15th International Conference on ITS Telecommunications—ITST 2017, Warsaw, Poland, 29–31 May 2017.
51. Sukuvaara, T.; Mäenpää, K.; Ylitalo, R.; Konttaniemi, H.; Petäjäjärvi, J.; Veskoniemi, J.; Autioniemi, M. Vehicular networking road weather information system tailored for arctic winter conditions. *Int. J. Commun. Netw. Inf. Secur.* **2015**, *7*, 60–68.
52. Jain, V.; Lapoehn, S.; Frankiewicz, T.; Hesse, T.; Gharba, M.; Gangakhedkar, S.; Ganesan, K.; Hanwen, C.; Eichinger, J.; Ramadam Ali, A.; et al. Prediction based framework for Vehicle Platooning using Vehicular Communications. In Proceedings of the 2017 IEEE Vehicular Networking Conference—VNC 2017, Turin, Italy, 27–29 November 2017. [CrossRef]
53. Sukuvaara, T.; Ylitalo, R.; Katz, M. IEEE 802.11p Based Vehicular Networking Operational Pilot Field Measurement. *IEEE J. Sel. Areas Commun.* **2013**, *31*, 409–417. [CrossRef]
54. Ameixieira, C.; Cardote, A.; Neves, F.; Meireles, R.; Sargento, S.; Coelho, L.; Afonso, J.; Areias, B.; Mota, E.; Costa, R.; et al. HarborNet: A real-world testbed for vehicular networks. *IEEE Commun. Mag.* **2014**, *52*, 108–114. [CrossRef]
55. Massow, K.; Radusch, I. A rapid prototyping environment for Cooperative Advanced Driver Assistance Systems. *J. Adv. Transport.* **2018**. [CrossRef]
56. Thompson, R. *All-in-One Urban Mapping Using V2X Communication*; University of Tennessee: Chattanooga, TN, USA, 2012; Available online: http://tsite.org/wp-content/uploads/2012/11/Presentation-3-%E2%80%93-All-In-One-Urban-Mapping-Using-V2x-Communication.pdf (accessed on 30 April 2019).
57. Lewis, B. V2X Test Beds for Faster, Cleaner, Safer Transportation. IoT Design, 2016. Available online: http://iotdesign.embedded-computing.com/articles/v2x-test-beds-for-faster-cleaner-safer-transportation (accessed on 30 April 2019).
58. Nissan, Savari and UC Berkeley Make V2X Testbed in Sunnyvale. Telematics Wire, 2016. Available online: https://www.telematicswire.net/vehicle-telematics-vehicle-information-technology-and-navigation/nissan-savari-and-uc-berkeley-make-v2x-testbed-in-sunnyvale/ (accessed on 30 April 2019).

59. *Connected and Automated Vehicle Program Plan*. Report of Virginia Department of Transportation. 2017. Available online: https://www.hrtpo.org/uploads/docs/061218%20A3%20-%20Release_Final_VDOT_CAV_Program_Plan_Fall_2017.pdf (accessed on 30 April 2019).
60. Evans, R. Transportation Research Center Begins Construction of New CAV Test Facility in Ohio. *Automotive Testing Technology Online Magazine*. 2018. Available online: https://www.automotivetestingtechnologyinternational.com/videos/transportation-research-center-begins-construction-of-new-cav-test-facility-in-ohio.html (accessed on 30 April 2019).
61. Spirent, Tata Elxsi V2X Automotive Test System Adopted by Researchers. GPS World, 2016. Available online: http://gpsworld.com/spirent-tata-elxsi-v2x-automotive-test-system-adopted-by-researchers (accessed on 30 April 2019).
62. *NI and Shanghai University Collaborate on a 5G Ultra-Reliable Low-Latency Testbed for V2X Communications*. National Instruments Press Release. 2018. Available online: http://www.ni.com/newsroom/release/ni-and-shanghai-university-collaborate-a-5g-ultra-reliable-low-latency-testbed-for-v2x-communications/en (accessed on 30 April 2019).
63. Vella, H. Guangzhou Becomes Test-Bed for Self-Driving Cars. *Techwire Asia*. 2018. Available online: https://techwireasia.com/2018/02/guangzhou-becomes-test-bed-self-driving-cars (accessed on 30 April 2019).
64. Ying, W. Shanghai Takes Another Step to Push Connected Car Road Tests. *China Daily*. 2018. Available online: https://www.chinadailyhk.com/articles/238/62/107/1519963996724.html (accessed on 30 April 2019).
65. AUtomotive Testbed for Reconfigurable and Optimized Radio Access (AURORA). Available online: http://rsl.ece.ubc.ca/aurora (accessed on 30 April 2019).
66. *Ottawa Launches Canada's First on-Street Test of an Autonomous Vehicle*. City of Ottawa Press Release. 2017. Available online: https://ottawa.ca/en/news/ottawa-launches-canadas-first-street-test-autonomous-vehicle (accessed on 30 April 2019).
67. *All Aboard the Future of Sustainable Transport*. Monash University Press Release. 2018. Available online: https://www.monash.edu/engineering/about-us/news-events/latest-news/articles/2018/all-aboard-the-future-of-sustainable-transport (accessed on 30 April 2019).
68. Emery, K. Driverless Car to Hit Perth Roads in 2019. *The West Australian*. 2018. Available online: https://thewest.com.au/news/perth/driverless-car-to-hit-perth-roads-in-2019-ng-b88957514z (accessed on 30 April 2019).
69. Smart Mobility Program. Available online: http://www.infinitus.eee.ntu.edu.sg/Programmes/SMP/Pages/home.aspx (accessed on 30 April 2019).
70. Lakrintis, A. Singapore Emerges as Autonomous Vehicles Test Bed. Strategy Analytics, 2016. Available online: https://www.strategyanalytics.com/strategy-analytics/blogs/automotive/autonomous-vehicles/autonomous-vehicles/2016/08/15/singapore-emerges-as-autonomous-vehicles-test-bed (accessed on 30 April 2019).
71. Hayashi, Y.; Memezawa, I.; Kantou, T.; Ohashi, S.; Takayama, K. Evaluation of Connected Vehicle Technology for Concept Proposal Using V2X Testbed. *SEI Tech. Rev.* **2017**, *85*, 10–14.
72. *Action Plan for Realizing Automated Driving*; Japanese Ministry of Economy, Trade and Industry Report; 2018. Available online: http://www.meti.go.jp/english/policy/mono_info_service/connected_industries/pdf/ad_v2.0_hokokusho.pdf (accessed on 30 April 2019).
73. M-City. Available online: mcity.umich.edu (accessed on 30 April 2019).
74. Goldberg, M. Uber Built a Miniature Fake City in Pittsburgh to Test Self-Driving Cars. The Drive, 2017. Available online: http://www.thedrive.com/tech/15241/uber-built-a-miniature-fake-city-in-pittsburgh-to-test-self-driving-cars (accessed on 30 April 2019).
75. Korea Partially Opens Test Bed Road for Autonomous Vehicles. *The Korea Herald: Yonhap News*. 2017. Available online: http://www.koreaherald.com/view.php?ud=20171106000459 (accessed on 30 April 2019).
76. Madrigal, A.C. Inside Waymo's Secret World for Training Self-Driving Cars. *The Atlantic*. 2017. Available online: https://www.theatlantic.com/technology/archive/2017/08/inside-waymos-secret-testing-and-simulation-facilities/537648/ (accessed on 30 April 2019).
77. Zala Zone Project. Available online: https://zalazone.hu/en (accessed on 30 April 2019).
78. CERMcity. Available online: https://www.atc-aldenhoven.de/en/new-urban-environment.html (accessed on 30 April 2019).

79. *Detailed Pilot Overview Report*. C-Roads Platform Report. 2017. Available online: https://www.c-roads.eu/fileadmin/user_upload/media/Dokumente/Detailed_pilot_overview_report_v1.0.pdf (accessed on 30 April 2019).
80. Eco-At Project. Available online: http://eco-at.info/ (accessed on 30 April 2019).
81. Cooperative ITS Corridor Project. Available online: http://c-its-korridor.de/?menuId=1&sp=en (accessed on 30 April 2019).
82. InterCor Project. Available online: http://intercor-project.eu (accessed on 30 April 2019).
83. SCOOP@F Project. Available online: http://www.scoop.developpement-durable.gouv.fr/-en (accessed on 30 April 2019).
84. InterCor UK Section: A2/M2 Connected Vehicle Corridor (A2/M2 CVC). Available online: http://intercor-project.eu/homepage/operations-united-kingdom (accessed on 30 April 2019).
85. NordicWay Project. Available online: http://vejdirektoratet.dk/EN/roadsector/Nordicway/NordicWay1/Pages/Default.aspx (accessed on 30 April 2019).
86. Olsen, E. *NordicWay2*. Report. 2017. Available online: http://vejdirektoratet.dk/EN/roadsector/Nordicway/Documents/D2.%20NordicWay2_Presentation_v2.pdf (accessed on 30 April 2019).
87. *Harmonised C-ITS Specifications for Europe*. C-Roads Platform Report. 2018. Available online: https://www.c-roads.eu/fileadmin/user_upload/media/Dokumente/Harmonised_specs_text.pdf (accessed on 30 April 2019).
88. C-Roads Pilots: Core Members. Available online: https://www.c-roads.eu/pilots/core-members.html (accessed on 30 May 2019).
89. Heineke, K.; Kampshoff, P.; Mkrtchyan, A.; Shao, E. Self-Driving Car Technology: When Will the Robots Hit the Road? McKinsey & Company, 2018. Available online: https://www.mckinsey.com/industries/automotive-and-assembly/our-insights/self-driving-car-technology-when-will-the-robots-hit-the-road (accessed on 30 April 2019).
90. Ben-Elia, E.; Di Pace, R.; Bifulco, G.N.; Shiftan, Y. The impact of travel information accuracy on route-choice. *Transp. Res. Part C Emerg. Technol.* **2013**, *26*, 146–159. [CrossRef]
91. *Evaluation and Assessment Plan*. C-Roads Platform Report. 2018. Available online: https://www.c-roads.eu/fileadmin/user_upload/media/Dokumente/C-Roads_WG3_Evaluation_and_Assessment_Plan_Final.pdf (accessed on 30 April 2019).
92. Harb, M.; Xiao, Y.; Circella, G.; Mokhtarian, P.L.; Walker, J.L. Projecting travelers into a world of self-driving vehicles: Estimating travel behavior implications via a naturalistic experiment. In Proceedings of the 97th Annual Meeting of the Transportation Research Board, Washington, DC, USA, 7–11 January 2018.

© 2019 by the authors. Licensee MDPI, Basel, Switzerland. This article is an open access article distributed under the terms and conditions of the Creative Commons Attribution (CC BY) license (http://creativecommons.org/licenses/by/4.0/).

Article

Congestion Control in V2V Safety Communication: Problem, Analysis, Approaches

Xiaofeng Liu [†] and Arunita Jaekel [*,†]

School of Computer Science, University of Windsor, Windsor, ON N9B 3P4, Canada; liu1ew@uwindsor.ca
* Correspondence: arunita@uwindsor.ca; Tel.: +1-519-253-3000 (ext. 2996)
† These authors contributed equally to this work.

Received: 26 January 2019; Accepted: 27 March 2019; Published: 13 May 2019

Abstract: The emergence of Vehicular Ad Hoc Networks (VANETs) is expected to be an important step toward achieving safety and efficiency in intelligent transportation systems (ITS). One important requirement of safety applications is that vehicles are able to communicate with neighboring vehicles, with very low latency and packet loss. The high mobility, unreliable channel quality and high message rates make this a challenging problem for VANETs. There have been significant research activities in recent years in the development of congestion control algorithms that ensure reliable delivery of safety messages in vehicle-to-vehicle (V2V) communication. In this paper, we present a comprehensive survey of congestion control approaches for VANET. We identify the relevant parameters and performance metrics that can be used to evaluate these approaches and analyze each approach based a number of factors such as the type of traffic, whether it is proactive or reactive, and the mechanism for controlling congestion. We conclude this paper with some additional considerations for designing V2V communication protocols and interesting and open research problems and directions for future work.

Keywords: congestion control; V2V; VANET

1. Introduction

Without doubt, vehicular safety is an important problem in modern society. Each year, nearly 1.25 million people die in road crashes, on average 3287 deaths a day [1]. Over the years, various technologies have emerged in the vehicular traffic sector to help reduce accidents. Since 2001, when Vehicle Ad Hoc Network (VANET) was first introduced in [2] as a term, it has been widely perceived by government, car manufacturing industries, and academia as a promising concept for future realization of Intelligent Transportation System (ITS) thereby achieving safety and efficiency in our nearly overcrowded motorways [3]. To date, a number of standards have been implemented to accommodate the Vehicle-to-Vehicle (V2V) and Vehicle-to-Infrastructure (V2I) communications for safety-related applications. Short to medium range wireless communication known as Dedicated Short Range Communication (DSRC) [4] has been proposed for such vehicular communication. In the United States, the Federal Communication Commission (FCC) allocated 75 MHz spectrum for DSRC. Collectively the IEEE 1609 family, IEEE 802.11p and the Society of Automotive Engineers (SAE) J2735 form the key parts of the currently proposed Wireless Access in Vehicular Environment (WAVE) protocol stack [5], which provides the protocols from physical level to application level for vehicular communication. Although several safety-related standards (e.g., channel spectrum, safety message format) have been defined, there remain many aspects of DSRC performance requirements, e.g., defining sending rate, transmit power control, adaptive message rate control that need to be investigated thoroughly. There are significant DSRC-related social and technical challenges that have to be dealt with before large-scale deployment.

A key requirement of V2V communication is the reliable delivery of safety messages. These messages are typically broadcast to neighboring vehicles using DSRC/WAVE technology, based on CSMA/CA in the media access layer. Due to the multi-access wireless channel, many factors can cause the delay or failure of the dissemination of safety messages. The limited channel bandwidth available to transmit the safety messages means that the shared radio channels can become easily congested as vehicle density increases. In [6], experiments have demonstrated that channel congestion can occur even in relatively simple traffic scenarios. Channel congestion is a critical factor that leads to delayed or failed messages delivery. With higher vehicle density, it is not clear if the channel capacity will be sufficient to support the data load generated by both beacons and event-driven safety messages. Therefore, the development of effective congestion control strategies for V2V communication is of utmost importance and has been an area of intense research interest in recent years.

Congestion control techniques for traditional wired and wireless communication has been well investigated in the literature [7–9]. For example, TCP [10] handles congestion with end-to-end control by adjusting the data rate on the source node, when detecting a change in the acknowledgments (ACKs) received from the destination node. In Mobile Ad hoc Networks (MANETs), researchers focus on routing and backward compatibility in multi-hop networks [11]. However, the unique challenges and requirements of VANET mean that these existing techniques are typically not suitable to be applied directly for VANET communication. For example, since safety messages are broadcast to all neighboring vehicles, collecting the ACKs from all the receiving vehicles becomes infeasible, as these ACK messages will consume bandwidth and further exacerbate the channel congestion. The single-hop based safety messages broadcasting in VANET also has completely different requirements compared to the multi-hop communication in MANETs. Furthermore, congestion control in VANET becomes even more challenging due to the high node mobility and channel fading constraints.

This paper will give a comprehensive overview of the requirements, performance metrics and current approaches for designing a congestion/awareness control protocol for V2V communication. The remainder of the paper is organized as follows. In Section 2, we introduce the relevant background information on V2V safety messages communication using DSRC/WAVE and the effects of higher channel load on such communication. In Section 3 we identify some of the relevant parameters and performance metrics for evaluating congestion control algorithms in VANET. In Section 4, we analyze and discuss the different congestion control protocols for VANET available in the literature and present our conclusions along with some open research problems and directions for future work in Section 5.

2. Background

Vehicular safety means not only strengthening the body of the vehicle to withstand crashes, but also improving a vehicle's awareness of surrounding vehicles and vice versa. Currently available camera-based safety techniques work within a limited distance, limited scope and require good visibility. Inter-vehicle wireless communication can help to overcome many of these limitations. The prompt delivery of safety messages, e.g., velocity, position, can be very helpful collision avoidance and other emergency situations. Since the 1920s, radio communications between vehicles have been used to improve safety [12]. Today, a wide variety of safety applications can be developed based on ITS technologies. One popular method to classify these applications is Time-To-Collision (TTC) [13], which measures the traffic conflict probability as the time required for two vehicles to collide if they continue at their present speed and on the same path [14]. In [15], authors argue that one of the congestion control principles should be deference to safety applications' requirements in different driving contexts. This means the applications will disseminate safety messages in different ways based on the vehicle's driving context.

In V2V communication safety-related messages are disseminated by *broadcasting* and consist of two main messages by two main types—(i) periodic messages and (ii) event-driven messages. The *Basic Safety Messages* (BSMs) defined by SAE J2735 [16] in the U.S. and *Cooperative Awareness Messages* (CAMs) specified by ETSI TC ITS [17] in Europe, also called as *beacons*, are transmitted *periodically* to announce the vehicle status. The *event-driven* messages are transmitted at the detection of a traffic event or road hazard, e.g., Decentralized Environmental Notification Message (DENM) developed by ETSI TC ITS [18]. In this paper, we primarily focus on BSMs for periodic broadcasts. BSMs consist of the following two parts:

- Part I includes basic vehicle state which is mandatory in each BSM, e.g., time, position, motion, vehicle size.
- Part II includes vehicle safety extension, which is optional for V2V safety applications, e.g., event flags, path history.

The average size of a BSM is from 320 to 350 Bytes, but if security-related overheads are considered, it could increase to around 800 Bytes. The default transmit rate is 10 times per second.

2.1. DSRC Spectrum for V2V Communication

FCC has allocated a 75 MHz spectrum in the 5.850–5.925 GHz (5.9 GHz band) divided into seven 10 MHz channels, with a 5 MHz guard band at the low end, for DSRC communication. In this spectrum, Channel 172 is designated for safety [19] and vehicle safety messages are expected to be exchanged on this channel. In [4], the authors propose that pairs of 10 MHz channels (channel 172 and 174) can also be combined into a 20 MHZ channel, to achieve increased capacity with shorter transmitting time, thus reducing the collision probability for the frame transmission. But a 20 MHz channel has been reported to lead to higher inter-symbol interference (ISI) and inter-carrier interference (ICI) that can significantly affect the successful reception of packets [20]. So far, the testing of DSRC and other related research mainly focus on 10 MHz channels.

Based on orthogonal frequency division multiplexing (OFDM) and different modulation techniques, DSRC provides 8 different data transfer rates (3, 4.5, 6, 9, 12, 18, 24, 27 Mbps). The higher data rate, the higher is the signal-to-interference-plus-noise ratio (SINR) threshold (dB) required for packet reception. There has been only limited research work reported on how to optimally select the data rate for safety communications. In [21], the authors proposed a 6 Mbps data rate after analyzing the work in [22], which concluded the lowest probability of collision is with 24 Mbps data rate in the channel under low overall channel load and without considering channel fading. A 6 Mbps data rate has been widely adopted in many congestion control simulations in VANET [4,11,15,23], as well as in some standardization activities [24]. Other data rates have also been considered, e.g., 3 Mbps is used in [25–27] for its lowest SINR requirement.

Figure 1 shows the current DSRC channel band and how BSM is transmitted on channel 172. Each vehicle will run some safety applications, e.g., emergency electronic brake lights (EEBL), left turn assist (LTA). These safety applications will need to disseminate the vehicle's BSMs periodically and acquire the other vehicles' BSMs to complete their assigned tasks. All these BSMs are transmitted on channel 172 plus some event-driven messages as well.

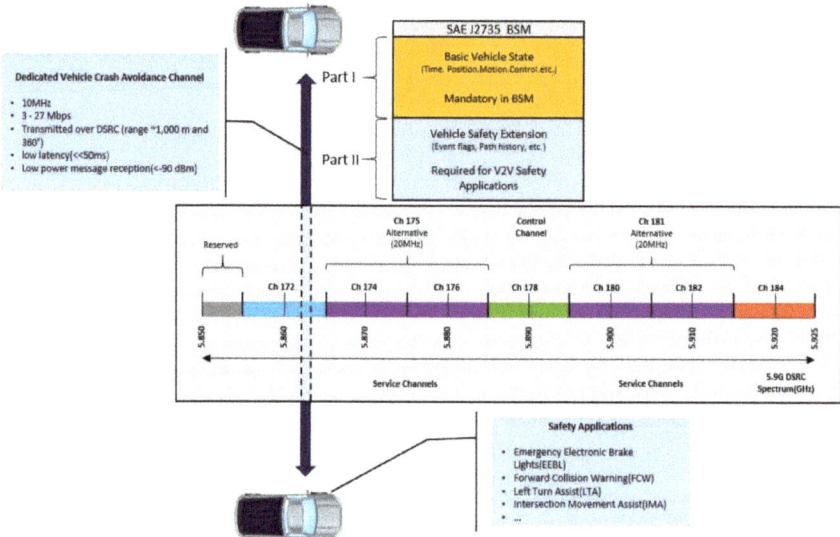

Figure 1. Dedicated Short Range Communication (DSRC) channel band and Basic Safety Messages (BSMs) transmission on channel 172.

2.2. Channel Congestion in DSRC

The 802.11p wireless protocol is currently the only protocol for low-latency V2V safety communication in VANET. All the safety messages are transmitted with 802.11p, which uses carrier sensing multiple access with collision avoidance (CSMA/CA) to reduce collisions and provide fair access to the channel. Due to the low latency requirements, the safety messages (both beacons and event-driven messages) are always transmitted using one-hop broadcasting communications. CSMA/CA is a contention-based random access protocol that can cause simultaneous transmissions, i.e., packet collisions, which significantly lower communication reliability [28,29]. For broadcast messages, re-transmission is usually not an option because receiving nodes do not send any acknowledgement (ACK) to avoid ACK flooding. Furthermore, the important packet-collision avoidance mechanisms like the request-to-send/clear-to-send (RTS/CTS) handshake are disabled in 802.11p MAC layer, resulting in hidden terminal problems.

The requirement to frequently broadcast BSMs can quickly lead to very high channel loads for the single 10 MHz channel that has been allocated for safety messages. For example, let us consider the following scenario: a 4-lane bidirectional highway with a vehicle density of 25 vehicle/km/lane, BSM transmissions every 100ms from each vehicle, where each BSM is 500 Bytes. The amount of data generated per second for mutual awareness could easily exceed the available data rate of 3 Mbps. Similar calculation are also given in [20,27]. The work in [14] lists three important consequences of high channel load for V2V communication:

- successful packet receptions decrease and transmission delay increases with increasing message generation rates;
- effective transmission ranges decrease under high channel load conditions;
- communication range of a transmitter decreases under interference (lower SINR) due to hidden stations or simultaneous sending.

The authors in [14] also identify the following reasons for packet loss under high channel load, which need to be considered in the design of congestion control protocols:

- transmission collision due to simultaneous sending;
- lower SINR caused by the interference from single/multiple hidden stations;
- packet dropped locally due to the failure of medium access;
- lower SINR caused by the nearby transmitting stations.

Both the MAC transmission delay and the number of packet collisions have been shown to grow rapidly when the channel load increases over a threshold of the channel access capacity, e.g., above a channel busy ratio (CBR) of 40% [14]. The MAC transmission delay will cause safety messages to arrive late, and a high number of packet collisions lead to a lower reception probability and hence an effective radio range reduction. When this occurs, it is referred to as *channel congestion*. In order to ensure reliable and timely delivery of safety messages, effective countermeasures need to be taken. In the following sections, we well discuss and compare some of the important congestion control strategies and their objectives for V2V communication.

3. Performance Metrics and Simulation Parameters

A number of different strategies for controlling congestion, in the allocated wireless channel for V2V communication, have been proposed in recent years. However, there is a wide divergence in terms of the objectives, performance criteria, simulation scenarios and other characteristics associated with these approaches, as reported in the literature. In order to have a meaningful comparison of the various approaches, we need to first identify the relevant metrics used to evaluate them, as well as the specific scenarios and channel characteristics, for which each approach was evaluated. In this section, we first consider the metrics that have been used to evaluate to performance of V2V congestion control algorithms. We then discuss the some of the different scenarios, e.g., traffic patterns, channel characteristics that affect the evaluation of these algorithms through simulations.

3.1. Performance Metrics

There are many different performance metrics for evaluating congestion control algorithms. Some of the most commonly used ones for V2V communication are discussed below.

3.1.1. Channel Busy Ratio (CBR)

CBR (Channel Busy Ratio) or CBT (Channel Busy Time) is defined as ratio between the time the channel is sensed as busy and the total observation time (e.g., 100 ms). It is a measure for the channel load perceived by a vehicle, and depends on the number of vehicles in its transmission range and their individual message generation rates. Many congestion control algorithms, such as periodically updated load sensitive adaptive rate control (PULSAR) [15], random transmit power control (RTPC) [30], distributed fair transmit power adjustment (D-FPAV) [25], distributed network utility maximization (D-NUM) [31], packet-count based decentralized data-rate congestion control algorithm (PDR-DCC) [32] use CBR/CBT as one of the metrics to analyze the communication benefit of the algorithms. CBR has been shown to be a suitable metric to increase packet delivery performance [33]. Simulations show that high message rates cannot be supported for all vehicles in case of high vehicle densities or lager packets [14]. In [34], the authors investigate the relationships between the CBR locally perceived by vehicles, i.e., the one-hop CBR, and the two-hops CBR. CBR can also be an input to congestion control mechanism. CBR has been referred to in other documents as the *Channel Busy Function (CBF)* [11] and the *Channel Busy Percentage (CBP)* in (V2V Safety Communication Scalability based on the SAE J2945/1 Standard). For analytical research about CBR, paper [35] gives a mathematical model in terms of CBR for channel load at a single position, load contribution from a single transmitter and channel load distribution on road respectively. Paper [36] introduces an analytical model based on a straightforward Markov reward chain to obtain transient measurements of the idle time of the link between two VANET nodes.

3.1.2. Packet Loss Rate (PLR)

Packet loss may be caused by different reasons, e.g., collisions, fading or control channel interval (CCHI) expiry time. PLR measures the percentage of packets lost as a result of collisions and/or fading. The percentage of dropped packets due to the CCHI expiry time is the ratio of dropped packets to the total packets generated by the application layer [37]. There are other similar metrics like PLR used in literature. *Packet error rate* is used in [38]. *Packet collision rate*, which is defined as the number of packet collisions normalized in time and space, is used in [30]. *Reception probability* is used in [21,25]. The inverse of PLR is *packet delivery ratio* (PDR), the analytical expression for PDR is given in [39].

3.1.3. Inter-Packet Delay (IPD)

The *inter-packet delay (IPD)* is the average delay between successive packets received from surrounding vehicles. It is also referred to as packet *inter-reception time (IRT)* in [40], which is the time between two sequential packet receptions for a particular sender-receiver pair, and this metric is used to assess the temporal aspect of awareness in each transmission power control algorithm. A higher IRT value means older information is being received from the neighboring vehicles.

Both PLR and IPD are important metrics for VANET, as they both give a measure of the level of awareness of surrounding vehicles. As such, both metrics have been widely used in performance evaluation of V2V communication protocols [21,25,30,37–40]. Although both PLR and IPD provide information on vehicular awareness, they give different perspectives on awareness. PLR is used to estimate what percentage of packets are reaching a vehicle, which gives a coarse-grained measure of vehicular awareness. On the other hand, IPD can provide a more in-depth view, e.g., difference in awareness for cases where the PLR may be the same. This is illustrated using the two scenarios in [41], where 50% of the beacons transmitted by vehicle A with a 10Hz frequency in an interval of 10 s are received at vehicle B, i.e., PLR = 50% in both cases.

- Scenario 1: Alternate beacons are received correctly. In this case IPD = 200 ms, meaning information for vehicle A is outdated for at most 200 ms;
- Scenario 2: Beacons are received in batches—25 beacons are received in the first two seconds of the interval, then no beacon is received for 5 s, and the remaining 25 beacons are received in the last three seconds of the interval. In this case, there is a situation awareness blackout of at least 5 s. Considering that a vehicle position can change by over 100 m in 5 s at highway speeds, it is clear that situation awareness is severely impaired in Scenario 2, resulting in possibly undetected dangerous situations.

3.1.4. Additional Metrics

Besides the aforementioned metrics, some other metrics have also been used to evaluate the performance of congestion control techniques. In [27] the *probability of successful reception of beacon message* is used with respect to the distance to assess the algorithm's effectiveness and the average *channel access time (CAT)* for claiming fairness. *Update delay* is defined in [30] as the elapsed time between two consecutive CAMs/BSMs successfully received from the same transmitter, which aims to assess the awareness of vehicle from a communication perspective. In [26] a *95% Euclidean cut-off error* (in meters) is defined and a *95% Tracking Error* in [23] is used to represent the tracking accuracy of the communication policy, which possesses a statistical sense similar to a *confidence interval (CI)*. The *information dissemination rate (IDR)*, i.e., the number of packets received successfully by a node's neighbours, is used in [33] as a metric for application performance. The authors conclude that CBR is a suitable feedback metric for maximizing the number of IDR, with a recommended CBR target value of 0.65.

3.2. Simulation Parameters

Due to the high cost and safety factors, simulation is an effective and popular way to evaluate congestion control approaches. In this section, we discuss some the important simulation tools and parameters that can affect the simulation outcomes.

3.2.1. Traffic Scenarios

Channel congestion can happen both on the highway or in urban areas, when vehicle density reaches a certain threshold. According to [42], a typical density for a 3-lane per direction on the highway could be 25 vehicles/Km/lane, which will result in a 300 vehicles within a 1000 m communication range. Authors in [14] claim that more challenging scenarios may be found on highways, because a simple velocity-based rate adaptation would solve the problem in urban area. Under normal low-traffic conditions, vehicles float in *free* flow. Due to perturbations at junctions or on-ramps, the state changes to *synchronized* flow, where vehicles on all lanes in the same direction have nearly the same speed. Due to further slight distortions, severe road traffic congestion occurs, also referred to as wide moving jams. These transitions of traffic flow are most relevant for decentralized congestion control, as the traffic density is high while at the same time the velocity is also considerably high. Traffic perturbations may cause emergency braking and accidents, which should be mitigated with the aid of inter-vehicle communication. Most of the congestion control approaches focus on the highway scenarios. But still some researches consider the urban scenarios [34,43]. The traffic scenarios are either deterministic, i.e., the vehicles are equally spaced with distance d m, or randomly positioned according to a *Poisson* distribution of average density ρ vehicles/m.

3.2.2. Channel Fading Model

The carrier signals are always degraded during transmission, which can limit both the carrier sense range and transmission range, and also cause hidden terminal problem. Appropriate channel fading models should be used to simulate the realistic environment, e.g., the deterministic Two-Ray Ground model, the probabilistic Log-Normal Shadowing model, Rayleigh model and Nakagami model. Among them, Nakagami model [44] is the most used one. Rayleigh model is a special case of Nakagami model, where the Nakagami fading parameter $m = 1$ and models a rough Non-Line-Of-Sight scenario. For fading parameter $m > 1$, Nakagami models an increased Line-Of-Sight scenario. In [45], the authors give a empirical conclusion: the lower the m is, the longer is the transmission distance used in the simulation. For example, when $m = 1$, the model is suitable for the transmission distances higher than 150 m. Since longer distance causes more fading, in [27] the authors refer to Nakagami $m = 1$ as *severe* fading conditions, to Nakagami $m = 3$ as *medium* fading conditions, and to Nakagami $m = 5$ as *low* fading conditions. We note that the suitable configuration values for channel fading can be highly dependent on different simulation parameters such as vehicle density, distance, presence of obstacles etc. and different papers have used different values.

3.2.3. Simulation Tools

There are many network simulators that can be utilized to help researchers understand the behavior and performance of the networks and their protocols. Here we list three of the most popular open source network simulators currently being used: ns-2, ns-3, OMNet++.

1. ns-2: Network Simulator-2 (ns-2) [46] is an open source, discrete event network simulator for both wired and wireless networks. In ns-2, arbitrary network topologies can be defined that are composed of routers, links and shared media [47]. The physical activities of the network are processed and queued in the form of events, in a scheduled order. These events are then processed as per the scheduled time, which increases along with the processing of events. However, the simulation is not real time, it is considered virtual [48]. ns-2 was extended by [49] with: (a) node mobility, (b) a realistic physical layer with a radio propagation model, (c) radio network

interfaces, and (d) the IEEE 802.11 Medium Access Control (MAC) protocol using the distributed coordination function (DCF). After revised by [50], the resulting PHY is a full featured generic module capable of supporting any single channel frame based communications. The key features include cumulative signal to interference plus noise ratio (SINR) computation, preamble and physical layer convergence procedure (PLCP) header processing and capture, and frame body capture. The MAC now accurately models the basic IEEE 802.11 carrier sense multiple access with collision avoidance (CSMA/CA) mechanism, as required for credible simulation studies.

2. ns-3: The ns-3 project [51] started from mid 2006 and is still under development. The latest release is ns-3.29. It is treated as a replacement instead of upgrading of ns-2. It supports parallel simulation and emulation using sockets. ns-3 provides a realistic environment and its source code is well organized compared to ns-2 [52]. In ns-3, vehicle mobility and network communication are integrated through events. User-created event handlers can send network messages or alter vehicle mobility each time a network message is received and each time vehicle mobility is updated by the model. The authors in [53] implemented a straight highway model in ns-3 that manages vehicle mobility, while allowing for various user customizations. The revised ns-3 has two main classes: Highway and Vehicle. Vehicles are fully-functional ns-3 nodes that contained additional information regarding their current acceleration, velocity, and position. The Highway class uses Model and LaneChange objects attached to Vehicles to move vehicles based on IDM (Intelligent Driver Model [54]) and MOBIL lane change model (Minimizing Overall Braking Induced by Lane Changes [55]). In addition, Highway used ns-3 callbacks to enable simulation developers to take control of Vehicles based on network messages, overriding, if need be, the standard controls used in Highways. More improvements for ns-3 can be found in [56].

3. OMNet++: Unlike ns-2 and ns-3, OMNet++ supports more than network simulation. It can also be used for multiprocessors modelling, distributed hardware systems, etc. As a general discrete event, component-based open architecture simulation framework, OMNet++ can combine with SUMO (Simulation of Urban MObility) [57], a traffic simulator and Veins (Vehicles In Network Simulation) [58], which couples the network and traffic simulator to simulate the vehicle communication. With Veins each simulation is performed by executing two simulators in parallel: OMNeT++ (for network simulation) and SUMO (for road traffic simulation). Both simulators are connected via a TCP socket. The protocol for this communication has been standardized as the Traffic Control Interface (TraCI). This allows bidirectionally-coupled simulation of road traffic and network traffic. Movement of vehicles in the road traffic simulator SUMO is reflected as movement of nodes in an OMNeT++ simulation. Nodes can then interact with the running road traffic simulation.

Paper [59] compared the performance of these simulators with the ad hoc on demand distance vector (AODV) routing algorithm and concluded that ns-3 has the fastest performance in terms of computation time. Both ns-2 and ns-3 fully utilize the CPU, while OMNet++ can provide more flexible functions. All three simulators can satisfy most of the vehicle communication simulation needs. They all can set up the common simulation parameters, e.g., noise floor, carrier sense threshold, packet reception SINR, payload size, data rate, Tx rate and power. More details about the survey and comparison of the simulators can be found in [60–62].

Although simulation-based VANET research has already produced important results, it has to be considered that wireless simulators are not entirely accurate representations of the real life situation, e.g., simplified propagation and interference models, inaccurate application data patterns [38]. A validation of simulation results using implementations on actual hardware (and vice versa) is necessary in VANET research. There are several existing testbeds for studying various VANET based protocols and applications. DRIVE [63] (DemonstRator for Intelligent Vehicular Environments) is an experimental implementation featuring both advanced vehicular services and vehicular networking (i.e., intra-vehicle, V2V, V2I communications). It provides a modular, re-configurable architecture based on carPC, wireless sensors and rooftop antennae and uses wireless and mobile technologies, e.g., WiFi, UMTS and Bluetooth. In [38], the authors discussed how a wireless testbed could be suitable for VANET research by using only software adjustments, e.g., an approximation of the IEEE 802.11p standard using IEEE 802.11a hardware, the emulation of mobility based on link impairment and the use of low transmit power together with manual topology configuration. With these techniques, IBBT w-iLab.t can support high vehicle density, multi-hop and emulation of mobility in a three-floor building with some extensive facilities. In [64], SD-VANET (Software Defined Architecture for VANET) is proposed to solve the high complexity in terms of the control and management of the network infrastructure due to the coexistence of multiple different access technologies (e.g., already deployed Wi-Fi networks, DSRC and cellular network). The authors use software defined networking to enhance VANET with Wi-Fi access capability as both roadside units and vehicles. Additionally, a Wireless Access Management (WAM) protocol is proposed to provide wireless host management and basic flow admission with respect to the available bandwidth to validate the capability of the offered architecture. MDX VANET [65] is a prototype VANET network built on the Middlesex University Hendon Campus, London. It is used to investigate better propagation models, road-critical safety applications as well as algorithms for traffic management. Some researchers use testbeds to measure the beaconing performance for congestion control. In [41], the authors use two IEEE 802.11p compliant devices, namely the LinkBird-MX v3 units produced by NEC, to measure two important metrics for congestion control: beacon (packet) delivery rate (PDR), and the packet inter-reception (PIR) time. Authors in [66] implemented the congestion control (CC) protocol defined in (SAE) J2945/1 standard [67] using a congestion generation testbed of Remote Vehicles (RVs). The test uses a DSRC-enabled Host Vehicle (HV) to wirelessly transmit BSMs, a Ground Truth Equipment (GTE) mounted on the HV to accurately capture its position and a DSRC sniffer to remotely capture the BSMs received over-the-air (OTA). Furthermore, The authors used a Congestion Generation Tool (CGT) that can emulate up to 160 Remote Vehicles (RVs) in the setup (which could be expanded) and up to 80% Channel Busy Percentage (CBP) (i.e., a measure of channel occupancy) by transmitting a mix of BSM and WSM (WAVE Short Message) packets. For congestion control research with simulators, different comprehensive scenarios can be researched in a low-cost way, experiments can be repeatable and parameters (e.g., vehicle movement patterns, vehicle density) can be easily controlled. But simulation can't represent the real-life performance. Several studies [68–70] have discussed this problem in detail. Regarding signal propagation and interference, several aspects are often simplified in wireless network simulators. Besides networking aspects, other elements such as traffic patterns and end-user models can also be incomplete in simulations. The advantages of experimental implementations (i.e., testbeds) over the above-mentioned approaches are the gain of hands-on experience in close-to-real or real performance and behavioral issues, as well as favoring intuition shape practice. The drawback of Field Operational Tests (FOTs) is the practical limitations, e.g., difficulty of implementing repeatable experiments, high cost, specific traffic model requirements. More valid evaluation methods still need to be developed.

4. Analysis and Approaches

Many approaches have been proposed to tackle the congestion problem in VANET, using a range of different control mechanisms, parameters, metrics and problem definitions. In spite of this diversity of approaches, some common topics are generally considered when implementing congestion control techniques, e.g., the safety message type, transmission parameters, fairness. In this section, we will discuss some topics that congestion control measures need to consider and give an overview of the related the approaches. We will also give a comparison of the various approaches in terms of the criteria discussed in this paper.

4.1. Beacons (BSMs/CAMs) vs. Event-driven Safety Messages

As discussed in Section 2, beacons are used to periodically broadcast vehicle status information, while event-driven messages typically disseminate more urgent information. Therefore, ideally, event-driven safety messages should be delivered with shorter delay and lower collision probability, over a larger area. So, a higher transmission power is more suitable for event-driven safety messages. Beacons are normally more relevant to nearby vehicles, typically within a 150 m radius [26]. In [25], the authors note that there must be a strategy to balance event-driven messages and beacons and conclude that the amount of load resulting from beacons should be limited, to leave some available bandwidth for event-driven messages. Since beacons are transmitted by every vehicle on the road periodically, most congestion control approaches only consider the beacons as the messages transmitting on the channel [71]. In [72] the authors propose a V2V communication protocol specially for emergency warning dissemination, which is a common example of event-driven safety messages. Obviously both periodic beacons and event-driven messages should be considered in congestion control approaches. But due to the difference in priority and dissemination mode (one-hop or multi-hop), it is difficult to integrate the dissemination for both messages in one algorithm.

4.2. Reactive vs. Proactive

Congestion control mechanisms can be categorized as proactive (or adaptive in [14]) (i.e., feed forward) or reactive (i.e., feedback). Proactive approaches try to prevent congestion before it occurs; whereas reactive approaches try to handle congestion after it is detected. In [20], the authors use traditional control theory to depict the congestion control mechanism, as shown in Figure 2. The objective is usually determined based on the requirements of the safety application, e.g., a threshold channel load value. The controller is open- or closed-loop depending on whether it accepts the feedback from the result, which can be from the network or from another individual node. The controller will adjust the transmission parameters based on the traffic situation or scenario and the objectives.

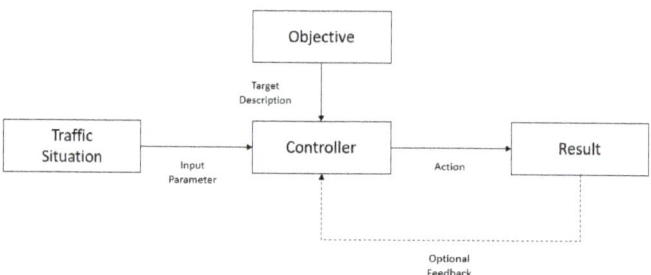

Figure 2. Congestion control mechanism from a control theory point of view [20].

Proactive approaches usually set up a target value for a specific of channel load metric, e.g., CBR and adapt the transmission parameters to drive the metric toward the target value. Then the algorithm can ensure the channel load will not exceed the target value. Some important proactive approaches include D-FPAV [27], linear message rate control algorithm (LIMERIC) [11], error model based adaptive rate control (EMBARC) [23], proactive ad hoc on-demand distance vector routing (Pro-AODV) [73], additive increase and mutliplicative decrease based decentralized congestion control (AIMD-Based DCC) [74], D-NUM [31] and RTPC [30].

An alternative approach is to react to congestion *after* it has actually occurred. Such a reactive approach is used in PULSAR [15], Adaptive transmission control [75], AC3 [76], etc.

Reactive strategy reacts directly to the measured channel load. The transmit behavior of can be pre-determined for any given channel load value. The function between channel load (e.g., CBR) and transmission parameters (e.g., Tx rate) determines the behavior and convergence of the control loop, when the number of nodes in range or their message rate changes. This freedom makes them very flexible, but analytically difficult to handle and understand. Additionally, since these algorithms react *after* congestion occurred, some BSMs may already get lost, which potentially can cause danger.

Proactive strategy can ensure that the channel load does not exceed a given threshold (e.g., CBR target value), but it needs to calculate the real-time information of the channel load and share with it's neighbours. If the shared information is too much and too frequent, the algorithm itself can contribute to channel congestion.

4.3. Selection of Control Parameters

Based on the channel capacity analysis theory in [77], when vehicle density increases, each node (i.e., vehicle) needs to either decrease the transmission *rate* (Tx rate) or reduce the transmission *power* (Tx power), so that the limited channel load can be shared by all the nodes. So, transmission power and rate are the two basic parameters that can be controlled to reduce channel congestion. Existing VANET congestion control techniques either adjust the *rate* of BSM packet transmission, or the transmit *power* or a combination of both to achieve acceptable levels of congestion.

4.3.1. Rate Control

The default Tx rate for BSMs/CAMs is 10 Hz, i.e., ten messages per second. At this rate, the channel load will exceed channel capacity when the vehicle density is high, leading to packet loss and delay. Rate control based approaches adapt the Tx rate to reduce the number of safety messages transmitted by each node per unit time, if necessary. The safety applications based BSMs/CAMs typically require a high probability information reception in a so called *target range*, i.e., 100 to 300 m [78]. In this context the inter-packet delay (IPD) is a common metric for application performance. Safety applications usually will ask for a minimum and a maximum value of IPD within the target range. Using *rate* control typically results in increased IPD, and may lead to unacceptable IPD values, if congestion is high. A number of recent congestion control algorithms use rate control, including PULSAR [15], which is based on the observation that the optimal choice of Tx rate depends on node density and the optimal Tx range does not. A similar binary message rate control approach is given in [79]. Other popular rate control approaches include LIMERIC [11], EMBARC [23] and a tracking error based approach presented in [26].

4.3.2. Power Control

Tx power determines how far a message can be sensed and be received successfully, i.e., the carrier sense (CS) range and the Tx range (normally 500 m with Tx power 10.21 dBm) respectively. Normally, the carrier sense range is longer (e.g., double) than the Tx range, which may decrease further due to channel fading and the interference from neighbours or hidden terminals. The goal of Tx power control is to adapt the Tx range, based on some performance metrics. Reducing the transmit power instead of transmission rate means that nearby vehicles will still see frequent BSMs. However, distant vehicles will not see any packets at all from a vehicle that is far away. So, power control sacrifices awareness of distant vehicles, but can maintain full awareness of nearby vehicles. Rate control, on the other hand, reduces awareness for both near and far vehicles; but, all vehicles within the transmission range will receive the BSMs. One of the most cited algorithms is D-FPAV [25], which uses a fully distributed, asynchronous and localized approach. Other power control based algorithms include SPAV [80], OSC [81], SBAPC [82], RTPC [83], NOPC [84] and AC3 [76], etc.

4.3.3. Hybrid Control

The ultimate objective for congestion control is to improve the awareness of the vehicles. Decreasing the Tx power or rate may mitigate congestion, but lowers the quality of awareness. Conversely, increasing the awareness by transmitting with maximum Tx power and rate increases congestion. As indicated in [85], Tx power and rate are inversely correlated at constant target load on the wireless channel, i.e., decreasing the Tx power allows an increase in the Tx rate and vice versa. The approaches discussed so far can only set one single power/rate pair, which results in the transmission power/rate trade-off dilemma [30], as shown in Figure 3. To fulfill the awareness range requirement (OP_{range}) by increasing the power, the rate has to be reduced, failing to achieve the required quality. On the other hand, to fulfill the quality requirement ($OP_{quality}$) by increasing the rate, the power has to be reduced, failing to achieve the required awareness range. To overcome this limitation, some joint power/rate congestion control approaches have been proposed. These approaches [30,75,86,87], attempt to jointly adjust rate and power to achieve the required level of awareness for all neighboring vehicles.

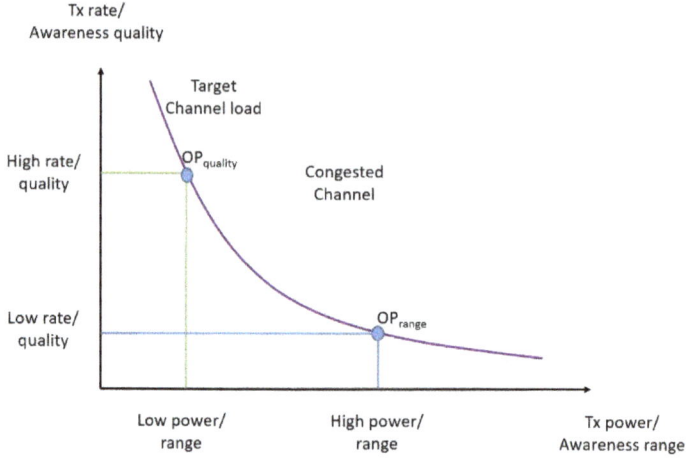

Figure 3. The trade-off dilemma by using constant transmit powers [30].

4.4. Overview of Different Congestion Control Approaches

In this section, we provide an overview of some of the leading congestion control algorithms in VANET. in terms of the parameters discussed in the previous sections. Table 1 lists the main characteristics of the leading congestion control approaches available in the literature.

PULSAR: PULSAR [15] (Periodically Updated Load Sensitive Adaptive Rate control) is a reactive approach, where each vehicle measures CBR at the end of a fixed time interval and compares the measured value against the target value. When the measured value is higher than the target value, i.e., congestion is observed, the vehicle will decrease its Tx rate accordingly. The PULSAR algorithm first fixes the Tx range based on the required target range and then adapts the Tx rate, with respect to CBR measurement. In PULSAR, the arrival of a new CBR measurement triggers a Tx rate adaption. The interval between CBR measurements is called *channel monitoring and decision interval* (CMDI) and is a fixed time for all nodes. When a new measurement arrives at the end of CMDI, PULSAR compares the measured value against the target value (CBR = 0.6) and adapts the Tx rate using Additive Increase Multiplicative Decrease (AIMD).

LIMERIC: LIMERIC [11] (Linear Message Rate Control algorithm) uses linear feedback to adapt the message rate instead of the limit cycle behavior inherent in PULSAR. In LIMERIC vehicles in a given region, i.e., a single collision domain, where all the nodes measure the same channel load, sense and adapt their Tx rates to meet a specified CBR. LIMERIC is a distributed algorithm, executed independently by each station in a neighborhood. If $r_j(t)$ is the message rate of the jth station at time t, and $CBR(t)$ is the CBR value measured at time t, then at each update interval LIMERIC adapts the message rate according to:

$$r_j(t+1) = (1-\alpha)r_j(t) + \beta(CBR_{target} - CBR(t)) \qquad (1)$$

In this equation, α and β are algorithm parameters that influence stability and adaptation speed. The final Tx rate is highly dependent on the number of vehicles in the region.

AIMD: In [74], the authors use the TCP's AIMD (Additive Increase and Multiplicative decrease) congestion control strategy, described in [88]. The AIMD rule is applied on the transmit rate r. Exceeding the channel load threshold C_{thresh} serves as decrease criterion. As long as the perceived channel load is below the threshold, the transmit rate is increased by an increment α, otherwise decreased with a factor β, where $0 < \beta < 1$. After the update function is evaluated, the resulting new transmit rate might be bounded by a predefined minimum and a maximum. If the CBR measurement interval is large or subject to averaging or smoothing, the CBR update might be slower than the transmit rate updates. Then a second decrease step is likely after a first decrease. This can be prevented by resetting the measured CBR.

MD-DCC: In [89], the authors propose a decentralized combined message-rate and data-rate congestion control (MD-DCC) scheme, which provides a fair and effective way of message-rate and data-rate allocation among vehicles to avoid congestion and satisfy application requirements. MD-DCC keeps the beacon frequency above the required minimum value by lowering the message-rate. However, the traffic density may reach a point where minimum frequency value is reached and traffic density still keeps increasing. At that point, MD-DCC increases the data-rate, resulting in more channel capacity. Adjusting data-rate for more channel bandwidth is a novel idea, but if the vehicles transmit message with different data-rates, the synchronization between the sender and receiver could be a big problem. Also higher data-rate needs higher SNR which is more difficult when vehicle density is getting higher.

EMBARC: EMBARC [23] (Error Model Based Adaptive Rate Control) extends the work of LIMERIC with the capability to preemptively schedule messages based on the vehicle's movement, i.e., a kind of suspected tracking error technique. This approach can provide extra transmit opportunities for highly dynamic vehicles, hence reduce incidences of large tracking error.

TRC for CASS: Another rate control algorithm TRC for CASS (Transmission Rate Control for Cooperative Active Safety System) considering tracking error is presented in [26], which focuses on the dissemination of real-time tracking information of neighbouring vehicles while avoiding network congestion and failure based on multiple scalar linear time-invariant (LTI) dynamical systems. The authors first evaluate the decentralized information dissemination policies for tracking multiple dynamical systems and use a Err-Coll-Dep [90] policy to get a better tracking performance, then calculate the transmission probability for each vehicle every 50 ms based on the suspected tracking error. Based on the transmission probability, the vehicle stochastically generates a packet with its latest state information and places it on the MAC queue for transmission. This approach does not transmit the safety message as a periodic beacon, but in a stochastic way based on tracking information instead of CBR.

D-FPAV: Paper [27] proposes a method called D-FPAV (Distributed Fair transmit Power Adjustment for Vehicular Ad Hoc Networks) to control the load of beacons and another method EMDV (Emergency Message Dissemination for Vehicular environments) for dissemination of event-driven messages. D-FPAV achieves congestion control by setting the node transmission power based on the prediction of application-layer traffic and the observed number of vehicles in the surrounding area. As a proactive approach, D-FPAV uses a predefined Max Beaconing Load (MBL) and calculates the network-wide optimal transmission power to keep the channel load under this value. By periodically gathering the information about AL (Application-layer Load) for nodes within maximum carrier sense range, a node first uses FPAV [71] to calculate the maximum Tx power for all the nodes within its maximum carrier sense range, without violating the MBL condition. Then the nodes exchange this power by broadcasting to the neighbours within the maximum carrier sense range. After getting the same message from the nodes in its vicinity, each node computes the final power level, which is the minimum value among the local maximum Tx power levels in its vicinity. In FPAV, all the nodes increase their Tx power simultaneously, by the same amount, as long as the MBL condition is satisfied.

The authors of D-FPAV also consider strategies to deal with event-driven messages, i.e., emergency messages to effectively balance both low-priority traffic (beacons), and high-priority messages. D-FPAV shows a clear prioritization of emergency over beacon messages and reception probability for both message types are significantly higher at close distances, compared to no congestion control being applied. In D-FPAV, a key problem is how to acquire the knowledge about the presence of other nodes in the maximum carriers sense range. The carriers sense range is normally larger than the transmission range. This means it is not possible to get this information of the vehicles located outside the transmission range with one hop broadcasting. D-FPAV provides this information by piggybacking the aggregated status information in only 1 out of 10 beacon messages, in order to reduce the overhead; but, the overhead can still grow very high.

SPAV: Paper [80] proposes a distributed algorithm SPAV (Segment-based Power Adjustment for Vehicular environments) that adjusts the Tx power, based on averaged position and communication values for the neighbour information instead of using detailed neighbour information about each single node in [27]. To get the minimum 'required knowledge' of positional information on neighboring nodes that is necessary to guarantee a preconfigured maximum beaconing load in the network, the authors propose a distributed vehicle density estimation (DVDE) strategy to provide an approximation of the surrounding traffic situation to each vehicle. This approximation is then locally used by a separate power control SPAV. SPAV is executed locally at each vehicle in order to compute a common maximum transmission power for all vehicles in its environment. Therefore, SPAV derives a so called catchment area, the maximum area in which vehicles are allowed to transmit beacons with a maximum common power only affecting the center of the segment without violating segment's MBL threshold, for each segment in its environment partition using the merged vehicle density histogram provided by DVDE.

RTPC: In [83], the authors propose to increase the Awareness Quality by randomly selecting the TX power for each CAM transmission and vehicle. Each vehicle controls its current TX power by using a certain probability distribution over a valid TX power interval. Using random TX power selection the radio propagation conditions are shifted with each transmission as well as the collision and interference areas. This mitigates the recurrence of collisions and interference, caused by the periodicity of CAM transmissions in combination with slow relative speeds. The random power selection mechanism provides statistical fairness as all vehicles use the same probability distribution and thus the same average TX power. Hence, local fairness is maintained as well.

OSC: In [81], the authors present a new algorithm to handle congestion by altering the value of transmission power. They propose a proactive approach, where transmit power of BSMs oscillate between two levels. A specified number of low-power packets are transmitted followed by a single high-power packet. This pattern is repeated until a change in network conditions (e.g., channel congestion) is detected. If necessary, a rate control algorithm such as LIMERIC can be combined with this method for further control.

SBAPC: SBAPC (Speed Based Adaptive Power Control) [82] presents a novel speed-based approach for controlling transmission power of BSMs. IN SBAPC, each vehicle dynamically adjusts the transmit power of BSM packets, depending on its current speed. The main idea for SBAPC is that time to collision (TTC) with neighboring vehicles decreases as the speed of the vehicle increases. Therefore, BSMs from vehicle at higher speeds should be received by more distant vehicles, and such vehicles should use a higher transmit power. The results indicate a slight decrease in CBR compared to OSC. The IPD is similar to OSC for vehicles within 150 m of the ego vehicle. Between 150 m–260 m SBAPC has lower IPD compared to OSC; however, the IPD increases quickly beyond 260 m.

NOPC: NOPC (Non-cooperative beacon Power Control) [84] proposes a distributed congestion control algorithm based on non-cooperative game theory to find the equilibrium point, i.e., Nash Equilibrium (NE). The authors consider the power control is a submodular game, hence has a greatest and a least NE. In NOPC, each player, i.e., each vehicle, should pay higher price at higher congestion for increasing its power in terms of twice differentiable pay-off function which is a function of CBR. Each vehicle measures its local CBR to update its BSM power based on gradient method and does not require to exchange the algorithm related information with its neighbors. NOPC algorithm also has a per vehicle parameter that every vehicle can adjust, without communicating it with other vehicles to meet its application requirement. The simulation results show that NOPC is more efficient in terms of bandwidth, it also has good stability and convergence to the unique NE with different initial values of power for vehicles.

AC3: In [76], the authors propose Adaptive transmit power Cooperative Congestion Control (AC3) to allow vehicles to select their transmit power autonomously with respect to their local channel congestion. AC3 uses the notion of marginal contributions of vehicles towards congestion and determines a fair power decrease for vehicles using a shapely value system model. This model requires the vehicles with the highest marginal contributions, i.e., the highest level of transmit power used by vehicles, to reduce the most transmit power and vice versa. Unlike [84], AC3 models cooperative congestion control approach on the principles of cooperative game theory in [91], which allows us to construct models of conflict and cooperation among players (i.e., vehicles) in order to maximize individual payoffs.

Adaptive Transmission Control: In [75], the authors adjust the BSM Tx rate using AIMD (additive increase multiplicative decrease), based on MAC blocking when congestion happens, e.g., the CBR exceeds a defined threshold. The authors observe that any combination of Tx rate and Tx power results in the same IDR vs CBR curve, which means CBR can be a suitable feedback for maximizing IDR. They propose a methodology which first fixes Tx rate, based on vehicle tracking performance, and then adapts Tx power based on CBR. One drawback is that the optimal choice of Tx power depends not only on the current Tx rate, but also on vehicle density, which is hard to estimate correctly.

TPRC: In [87], the authors propose a TPRC (Transmit Power/Rate Control) strategy by first analyzing the spatio-temporal characteristics of awareness and identifying the average or kth percentile of IRT at a target distance d_t, as a suitable choice to assess the safety benefit associated with a particular Tx parameter (power or rate) setup. To find the optimal Tx parameter combinations, the authors first set the Tx power to the optimal value for the current d_t of the individual vehicle, then adapt the Tx rate with TRC (Transmit Rate Control). When d_t changes, TPC (Transmit Power Control) is applied. The main advantage of the suggested joint optimization strategy is that it efficiently approximates the identified optimal Tx parameter configurations for a given channel load target and d_t, while reducing complexity due to its independence of vehicle density.

RTPC + TRC: Considering awareness as in [83], the authors in [30] propose a new awareness/congestion control strategy. The authors first use RTPC (Random Tx Power Control) featuring spatial awareness control to reduce the channel load and then combine RTPC with transmit rate control(TRC) by subsequently increasing the current Tx rate until the target load is reached. This approach is similar with TPRC [87], but relies on randomized transmit power to provide a heterogeneous awareness quality in space and to relax the strict transmission range to power mapping. RTPC is to find an optimal Tx power distribution that in turn represents an optimal behavior of the awareness quality in space instead of the optimal constant Tx power in TPRC.

CPRC: CPRC (Combined Power and Rate Control) [86] proposes a mechanism for the combination of power and rate adjustment in a single algorithm loop, unlike the the two-phase joint power/rate control approaches in [30,87,92]. CPRC's policy is a cooperative behavior, where nodes that are not directly involved in a potentially dangerous situation decrease their Tx power to allow nodes affected by dangerous situations to increase their sending rate. It allows applications to send periodic messages at higher rates, when required by contextual factors like speed and higher probability of collision, without having the network load exceed the predefined reliability-based threshold.

4.5. Comparison of Different Approaches

Table 1 provides a comparison of leading congestion control approaches discussed in this section, in terms of the parameters discussed in the previous sections.

Table 1. Classification of congestion control algorithms.

Algorithm	Rate/Power Control	Proactive/Reactive	Message Type	Metrics Used	Channel Fading Model	Simulator
LIMERIC [11]	Rate	Proactive	Beacon	CBR	Not mentioned	ns-2
PULSAR [15]	Rate	Reactive	Beacon	CBR	Reyleigh, Nakagami	ns-2
AIMD [74]	Rate	Proactive	Beacon	Reception rate, IPD, CBR	Not mentioned	own simulator
EMBARC [23]	Rate	Proactive	Beacon	Tracking error	Not mentioned	ns-2
TRC for CASS [26]	Rate	Proactive	Beacon	95% cutoff error	Rayleigh	OPNET
D-FPAV [25]	Power	Proactive	Beacon, Event messages	Probability of message reception	Nakagami	ns-2
SPAV [80]	Power	Proactive	Beacon	Beaconing load	Nakagami, Two Ray Gound,	ns-2
RTPC [83]	Power	Proactive	Beacon	Packet collision rate, CBR, Update delay	Not mentioned	ns-3
OSC [81]	Power	Proactive	Beacon	Beacon error rate	Not mentioned	OMNet++, Veins
SBAPC [82]	Power	Proactive	Beacon	Beacon error rate, CBR, IPD	Not mentioned	OMNet++, Veins
NOPC [84]	Power	Proactive	Beacon	CBR	Nakagami	OMNet++, SUMO
AC3 [76]	Power	Reactive	Beacon	CBR	Not mentioned	OMNet++, Veins
Adaptive transmission control [75]	Hybrid	Reactive	Beacon	95% cutoff error	Rayleigh	OPNET
TPRC [87]	Hybrid	Proactive	Beacon	IRT, CBR	Power-law	ns-2
RTPC+TRC [30]	Hybrid	Proactive	Beacon	Packet collision rate, CBR, Update delay	Not mentioned	ns-3
CPRC [86]	Hybrid	Proactive	Beacon	Channel Busy Time	Not mentioned	ns-2.31
MD-DCC [89]	Hybrid	Proactive	Beacon	Channel Busy Time	Not mentioned	ns-3, SUMO

Table 1 shows that there is a fair balance between rate control and power control algorithms, although there appears to be a growing consensus that a joint approach combining both will likely lead to better results in the future. The vast majority of the current approaches focus mainly on periodic beacons. There is a clear need to implement strategies that allow high priority event-driven messages to be accommodated, even with high channel loads. Finally, in terms of performance evaluations, a wide range of metrics have been used. However, some metrics such as CBR, IPD and packet loss seem to be emerging as widely-used performance indicators. There is a need to identify a set of standard metrics that can be used to evaluate and compare different approaches consistently.

5. Additional Design Considerations for V2V Congestion Control

5.1. Fairness

Fairness is important in vehicular networks to ensure that all vehicles in the network have similar opportunities to communicate with nearby nodes [20]. It implies that no vehicle should be allocated arbitrarily less resources than its neighbours. However, there is a trade-off between fairness and efficiency [93]: more fairness usually results in a less efficient use of the shared resource. Unfortunately, efficiency is critical for safety communication, so it is important to find a suitable balance between fairness and efficiency. Although many congestion control approaches consider fairness as an important objective, there is currently no universally used measure of fairness. This makes it difficult to evaluate and compare different approaches for fairness. However, in this section we highlight different ways in which some of the existing algorithms have considered fairness.

Fairness includes both local fairness and global fairness. Local fairness means nodes near each other share the same channel, i.e., they have similar congestion controls levels. In LIMERIC [11], *local* fairness is implemented among immediate neighboring vehicles by assuming all the vehicles are in a single collision domain. When the algorithm converges, all the vehicle will use the same Tx rate to transmit. The a concept of "fair power control", where all vehicles in a certain area must restrict their beacons' transmission power by the same ratio is introduced in [71]. A max-min principle is used to achieve fairness, which aims to find a per-node maximal power assignment and the minimum of the nodes' transmit power levels is maximized as long as network load conditions are satisfied. However, the max-min principle is very difficult to apply, due to the unbounded and probabilistic nature of vehicular wireless communication. Authors in [15] deem that it is not necessary to require nearby nodes to have the same Tx parameters.

For *global* fairness, all the vehicles contributing to congestion will be involved, which means all the vehicles in the carries sense (CS) range need to exchange the channel state information. The CS range is always larger than the Tx range, so one-hop is not sufficient to notify all the nodes within CS range. In [15], the authors propose a two-hop piggybacking to share the channel state with the CS range, hence to fulfill the global fairness principle. This maximizes the Tx rate of the nodes that are throttled the most by congestion control, while not unnecessarily constraining other vehicles which do not contribute to congestion. One drawback is that nodes measuring congestion would decrease their Tx rates first, while others contributing to congestion might not yet be aware of doing so, and increase their Tx rates even further. Effective ways to react to the same state of the system at the same time is still a critical problem for global fairness.

5.2. Awareness Control

The vast majority of the approaches discussed so far aim to control *congestion* by appropriately reducing the channel load. However, the ultimate objective of congestion control is to improve the overall vehicle *awareness*, hence to achieve the safety of the vehicles. With this view, the authors in [30] define *awareness* in the context of vehicular safety communication as follows:

Definition 1 (Awareness). *It is the ability of an application to know the status, e.g., position, speed, heading, of neighboring vehicles. Awareness is qualified by its range, i.e., distance at which the application at most becomes aware of vehicles, and its quality, i.e., accuracy/up-to-dateness of the status information.*

Most of the papers we have discussed so far focus on how to control the channel load by adapting the transmission parameters, so called congestion control protocols. The concept of *awareness control* is explicitly proposed in [20], which aims at ensuring each vehicle's capacity to detect as well as to communicate with the relevant vehicles in their local neighborhood. Awareness control focuses more on the satisfaction of its individual safety applications and the traffic information of relevant local neighborhood. The issues have also been considered in some of the congestion control protocols. Based on observation that congestion and awareness control protocols have been typically designed and evaluated separately, the authors propose INTERN [94], which dynamically adapts the Tx power and rate of beacons of each vehicle to satisfy its application requirements, while controlling the channel load as the same time. The authors also claim that awareness control can achieve both local and global fairness. But when different safety applications have different transmission requirements, it is very difficult to guarantee that all vehicles in the neighbourhood will have similar transmission parameters. Awareness control and congestion control can even be in conflict in some cases, e.g., when some emergency event happens. The author of [94] proposed COMPASS (cross-layer coordination of multiple vehicular protocols) in [95] which try to coordinate awareness control and congestion control protocols. This approach has to get the transmission parameters firstly from the protocols and find a intersection configuration set between awareness control and congestion control protocols which might fail.

To show how awareness considerations can influence the delivery of different safety messages in VANET, we will examine several typical scenarios.

In Figure 4a we assume a low vehicle density on the road, where each vehicle is sending the safety messages periodically in a radio range with some transmission power and rate. At this time, the channel load is under a "safe" level, usually 40% of the theoretical maximum channel capacity [14], which ensures that all the safety messages can be delivered to the other vehicles in the transmission range. As the vehicle density increases, as shown in Figure 4b, the channel load will exceed the "safe" level, which means MAC transmission delay and the number of packet collisions will grow rapidly. The MAC transmission delay will cause safety messages to arrive late, and a high number of packet collisions will lead to a lower reception probability and hence an effective radio range reduction. To decrease the channel load, all the vehicles have to take some countermeasure, i.e., congestion control, to reduce the number of safety message packets. Since there are no "special" events, fairness in this scenario implies that every vehicle should have the same chance to share its current state, i.e., same Tx rate and power.

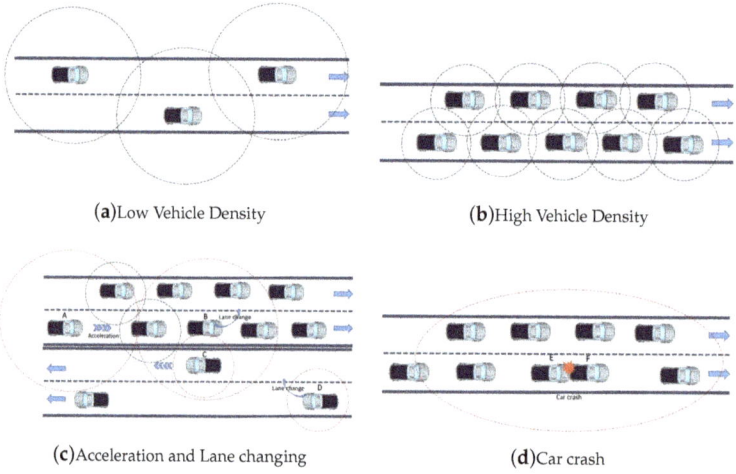

Figure 4. Driving Scenarios.

In Figure 4c we list two typical maneuvers: acceleration and lane changing. Vehicle A and B are driving in the higher density context, so the maneuvers should be noticed instantly by their neighbours to prevent an accident by enhancing the transmission power or rate. However, vehicle C and D have different driving context, they can finish the maneuvers without any special adjustment of the transmission parameters. In [20] the authors defined awareness control to differentiate it from congestion control. Awareness control aims at ensuring each vehicle's capacity to detect, and possibly communicate with the relevant vehicles and infrastructure nodes present in their local neighborhood as needed, through the dynamic adaptation of their transmission parameters. Since Vehicle A and B's safety applications require more channel resources than the nearby vehicles to maintain adequate awareness, they may use a higher Tx power, compared to the surrounding vehicles, as shown in Figure 4c. Finally, in Figure 4d a car crash is assumed. The safety message here is event-driven and should be delivered to the other vehicles at the largest radio range at one time.

There are still many challenging issues in implementing appropriate awareness control strategies, such as how to define the "relevant" vehicles for a particular safety application. For example, in Figure 4c, even if vehicles C and D are on the opposite lanes as A and B, their signals may still contribute to the congestion of the area where vehicle A and B are located. The concept of awareness control is an very important research problem for V2V safety communication. The goal is to find a suitable balance between a single vehicle's safety application's requirement, while maintaining low channel load and fairness among vehicles.

6. Conclusions

V2V safety communication enhances transportation safety, by improving the awareness of the vehicles on the road. The available 10 MHz channel cannot guarantee the delivery of safety messages without latency and error when vehicle density increases. The dynamic topology of vehicular networks makes the dissemination of safety messages even more difficult and error-prone. In this paper, we reviewed the congestion problem in V2V communication and discussed the important approaches for congestion control proposed in the literature. We analyzed how congestion occurs and discussed the most commonly used metrics when evaluating the channel performance. We also identified different simulation parameters that affect the evaluation of the algorithms. Finally, we classified and compared the leading congestion control approaches in terms of different relevant criteria, such as control mechanism, parameter being adjusted, and message type.

Congestion control in VANET is an important and challenging problem. Although, some interesting and effective strategies have been reported in the literature, there are still many challenges and open research problems for congestion control in V2V safety communication in the future. A few directions for future research are given below.

- Joint power/rate control: Existing hybrid approaches that combine Tx power and rate adaption typically implement congestion control in two different phases, e.g., fix a Tx power first and then adapt the Tx rate. A real time combined Tx power and rate control based on detailed safety benefit calculations can lead to improved performance and is a promising direction for research.
- Improved awareness control: As discussed in Section 5.2, awareness control focuses more on the relevant vehicles and local vehicles' information. We need to be able to accurately identify "relevant" vehicles and acquire detailed local traffic information, e.g., awareness needs more specific information about the vehicle's position, speed. So far only a few papers have considered the tracking error when implementing congestion control.
- Relative fairness: Each vehicle may have a very different driving context. So absolute fairness for both local or global fairness is not realistic. More specific relative fairness criteria are needed, when adapting the transmission parameters. It is an important problem to develop suitable metrics for evaluating fairness in different contexts and design approaches that maximize fairness.
- Standardization: In this paper, we reviewed many different metrics and approaches for V2V safety communication. It is not realistic to follow one unified process to deal with congestion problem, however at least the metrics used in the approaches should be normalized. For example, the widely used CBR metric has been referred to by a number of different names in different papers. There is a need for adoption of a common terminology and method of calculation for the performance metrics, to ensure consistent and fair evaluation of the various approaches.

Author Contributions: X.L., A.J. designed the paper, X.L. collected the data, performed the analysis, and wrote the paper under the supervision of A.J. A.J. supervised the research and provided guidance and key suggestions in writing the paper.

Funding: This research was funded by NSERC DG, Grant# RGPIN-2015-05641.

Acknowledgments: The work of A. Jaekel has been supported by a research grant from the Natural Sciences and Engineering Research Council of Canada (NSERC).

Conflicts of Interest: The researcher claims no conflict of interests.

References

1. World Health Organization. *Global Status Report on Road Safety 2015*; World Health Organization: Geneva, Switzerland, 2015.
2. Toh, C. *Ad Hoc Mobile Wireless Networks: Protocols and Systems*; Pearson Education: London, UK, 2001.

3. Eze, E.C.; Zhang, S.J.; Liu, E.J.; Eze, J.C. Advances in vehicular ad-hoc networks (VANETs): Challenges and road-map for future development. *Int. J. Autom. Comput.* **2016**, *13*, 1–18. [CrossRef]
4. Kenney, J.B. Dedicated Short-Range Communications (DSRC) Standards in the United States. *Proc. IEEE* **2011**, *99*, 1162–1182. [CrossRef]
5. Li, Y. *An Overview of the DSRC/WAVE Technology*; Springer: Berlin/Heidelberg, Germany, 2012; Volume 74. [CrossRef]
6. Weinfeld, A. Methods to reduce DSRC channel congestion and improve V2V communication reliability. In Proceedings of the 17th ITS World Congress, Busan, Korea, 25–29 October 2010.
7. Paganini, F.; Doyle, J.; Low, S. Scalable laws for stable network congestion control. In Proceedings of the 40th IEEE Conference on Decision and Control, Orlando, FL, USA, 4–7 December 2001; Volume 1, pp. 185–190.
8. Yi, Y.; Shakkottai, S. Hop-by-hop congestion control over a wireless multi-hop network. *IEEE/ACM Trans. Netw.* **2007**, *15*, 133–144. [CrossRef]
9. Welzl, M. *Network Congestion Control: Managing Internet Traffic*; John Wiley & Sons: Hoboken, NJ, USA, 2005.
10. Allman, M.; Paxson, V.; Blanton, E. TCP Congestion Control; No. RFC 5681; 2009. Available online: https://www.rfc-editor.org/rfc/rfc5681.txt (accessed on 10 May 2019).
11. Bansal, G.; Kenney, J.B.; Rohrs, C.E. LIMERIC: A linear adaptive message rate algorithm for DSRC congestion control. *IEEE Trans. Veh. Technol.* **2013**, *62*, 4182–4197. [CrossRef]
12. Flurscheim, H. Radio Warning Systems for use on Vehicles. US Patent 1,612,427, 13 November 1925.
13. Hayward, J.C. Near Miss Determination through Use of a Scale of Danger. Pennsylvania Transportation and Traffic Safety Center. 1972. Available online: https://onlinepubs.trb.org/Onlinepubs/hrr/1972/384/384-004.pdf (accessed on 10 May 2019).
14. Campolo, C.; Molinaro, A.; Scopigno, R. *Vehicular Ad Hoc Networks-Standards, Solutions, and Research*; Springer: Berlin/Heidelberg, Germany, 2015.
15. Tielert, T.; Jiang, D.; Chen, Q.; Delgrossi, L.; Hartenstein, H. Design methodology and evaluation of rate adaptation based congestion control for vehicle safety communications. In Proceedings of the 2011 IEEE Vehicular Networking Conference (VNC), Amsterdam, The Netherlands, 14–16 November 2011; pp. 116–123.
16. *SAE J2735: Dedicated Short Range Communications (DSRC) Message Set Dictionary*; Society of Automotive Engineers, DSRC Committee: Warrendale, PA, USA, 2009.
17. ETSI (2013) ETSI EN 302 637-2 (V1.3.0)—Intelligent Transport Systems (ITS); Vehicular Communications; Basic Set of Applications; Part 2: Specification of Cooperative Awareness Basic Service; Vehicular Communications. 2013. Available online: https://www.etsi.org/deliver/etsi_en/302600_302699/30263702/01.03.02_60/en_30263702v010302p.pdf (accessed on 10 May 2019).
18. ETSI (2013) ETSI EN 302 637-3 V1.2.0—Intelligent Transport Systems (ITS); Vehicular Communications; Basic Set of Applications; Part 3: Specification of Decentralized Environmental Notification Basic Service; Vehicular Communications. 2013. Available online: https://www.etsi.org/deliver/etsi_en/302600_302699/30263703/01.02.00_20/en_30263703v010200a.pdf (accessed on 10 May 2019).
19. *Amendment of the Commission's Rules Regarding Dedicated Short-Range Communication Services in the 5.850–5.925 GHz Band (5.9 GHz Band)*; Technical Report; U.S. Federal Communications Commission: Washington, DC, USA, 2006.
20. Sepulcre, M.; Mittag, J.; Santi, P.; Hartenstein, H.; Gozalvez, J. Congestion and Awareness Control in Cooperative Vehicular Systems. *Proc. IEEE* **2011**, *99*, 1260–1279. [CrossRef]
21. Jiang, D.; Chen, Q.; Delgrossi, L. Optimal data rate selection for vehicle safety communications. In Proceedings of the Fifth ACM international workshop on VehiculAr Inter-NETworking, San Francisco, CA, USA, 15 September 2008; pp. 30–38.
22. Xu, Q.; Mak, T.; Ko, J.; Sengupta, R. Vehicle-to-vehicle safety messaging in DSRC. In Proceedings of the 1st ACM International Workshop on Vehicular Ad Hoc Networks, Philadelphia, PA, USA, 1 October 2004; pp. 19–28.
23. Bansal, G.; Lu, H.; Kenney, J.B.; Poellabauer, C. EMBARC: Error model based adaptive rate control for vehicle-to-vehicle communications. In Proceedings of the Tenth ACM International Workshop on Vehicular Inter-Networking, Systems, and Applications, Taipei, Taiwan, 25 June 2013; pp. 41–50.
24. ETSI. Intelligent Transport Systems (ITS); European Profile Standard on the Physical and Medium Access Layer of 5 GHz ITS. 2009. Available online: https://www.etsi.org/deliver/etsi_es/202600_202699/202663/01.01.00_50/es_202663v010100m.pdf (accessed on 20 December 2018).

25. Torrent-Moreno, M.; Santi, P.; Hartenstein, H. Distributed fair transmit power adjustment for vehicular ad hoc networks. In Proceedings of the 2006 3rd Annual IEEE Communications Society on Sensor and Ad Hoc Communications and Networks, Reston, VA, USA, 28 September 2006; Volume 2, pp. 479–488.
26. Huang, C.L.; Fallah, Y.P.; Sengupta, R.; Krishnan, H. Intervehicle transmission rate control for cooperative active safety system. *IEEE Trans. Intell. Transp. Syst.* **2011**, *12*, 645–658. [CrossRef]
27. Torrent-Moreno, M.; Mittag, J.; Santi, P.; Hartenstein, H. Vehicle-to-vehicle communication: Fair transmit power control for safety-critical information. *IEEE Trans. Veh. Technol.* **2009**, *58*, 3684–3703. [CrossRef]
28. Sjöberg Bilstrup, K.; Uhlemann, E.; Ström, E.G. Scalability issues of the MAC methods STDMA and CSMA of IEEE 802.11 p when used in VANETs. In Proceedings of the 2010 IEEE International Conference on Communications Workshops, Cape Town, South Africa, 23–27 May 2010.
29. Kloiber, B.; Strang, T.; Röckl, M.; de Ponte-Müller, F. Performance of CAM based safety applications using ITS-G5A MAC in high dense scenarios. In Proceedings of the 2011 IEEE Intelligent Vehicles Symposium (IV), Baden, Germany, 5–9 June 2011; pp. 654–660.
30. Kloiber, B.; Harri, J.; Strang, T.; Sand, S.; Garcia, C.R. Random transmit power control for DSRC and its application to cooperative safety. *IEEE Trans. Dependable Secur. Comput.* **2016**, *13*, 18–31. [CrossRef]
31. Zhang, L.; Valaee, S. Congestion control for vehicular networks with safety-awareness. *IEEE/ACM Trans. Netw.* **2016**, *24*, 3290–3299. [CrossRef]
32. Math, C.B.; Li, H.; de Groot, S.H.; Niemegeers, I. Fair decentralized data-rate congestion control for V2V communications. In Proceedings of the 2017 24th International Conference on. IEEE Telecommunications (ICT), Limassol, Cyprus, 3–5 May 2017; pp. 1–7.
33. Fallah, Y.P.; Huang, C.L.; Sengupta, R.; Krishnan, H. Analysis of information dissemination in vehicular ad-hoc networks with application to cooperative vehicle safety systems. *IEEE Trans. Veh. Technol.* **2011**, *60*, 233–247. [CrossRef]
34. Autolitano, A.; Reineri, M.; Scopigno, R.M.; Campolo, C.; Molinaro, A. Understanding the channel busy ratio metrics for decentralized congestion control in VANETs. In Proceedings of the 2014 International Conference on IEEE Connected Vehicles and Expo (ICCVE), Vienna, Austria, 3–7 November 2014; pp. 717–722.
35. Chen, Q.; Jiang, D.; Tielert, T.; Delgrossi, L. Mathematical modeling of channel load in vehicle safety communications. In Proceedings of the 2011 IEEE Vehicular Technology Conference (VTC Fall), San Francisco, CA, USA, 5–8 September 2011; pp. 1–5.
36. Martin-Faus, I.V.; Urquiza-Aguiar, L.; Igartua, M.A.; Guérin-Lassous, I. Transient analysis of idle time in VANETs using Markov-reward models. *IEEE Trans. Veh. Technol.* **2018**, *67*, 2833–2847. [CrossRef]
37. Javed, M.A.; Khan, J.Y. A Cooperative Safety Zone Approach to Enhance the Performance of VANET Applications. In Proceedings of the VTC Spring, Dresden, Germany, 2–5 June 2013; pp. 1–5.
38. Vandenberghe, W.; Moerman, I.; Demeester, P.; Cappelle, H. Suitability of the wireless testbed w-iLab. t for VANET research. In Proceedings of the 2011 18th IEEE Symposium on Communications and Vehicular Technology in the Benelux (SCVT), Ghent, Belgium, 22–23 November 2011; pp. 1–6.
39. Killat, M.; Hartenstein, H. An empirical model for probability of packet reception in vehicular ad hoc networks. *EURASIP J. Wirel. Commun. Netw.* **2009**, *2009*, 4. [CrossRef]
40. Chang, H.; Song, Y.E.; Kim, H.; Jung, H. Distributed transmission power control for communication congestion control and awareness enhancement in VANETs. *PLoS ONE* **2018**, *13*, e0203261. [CrossRef]
41. Martelli, F.; Renda, M.E.; Resta, G.; Santi, P. A measurement-based study of beaconing performance in IEEE 802.11 p vehicular networks. In Proceedings of the 2012 IEEE INFOCOM, Orlando, FL, USA, 25–30 March 2012; pp. 1503–1511.
42. Manual, H.C. *Highway Capacity Manual*; Transportation Research Board: Washington, DC, USA, 2000; Volume 2.
43. Cheng, L.; Henty, B.E.; Stancil, D.D.; Bai, F.; Mudalige, P. Mobile vehicle-to-vehicle narrow-band channel measurement and characterization of the 5.9 GHz dedicated short range communication (DSRC) frequency band. *IEEE J. Sel. Areas Commun.* **2007**, *25*, 1501–1516. [CrossRef]
44. Nakagami, M. The m-distribution—A general formula of intensity distribution of rapid fading. In *Statistical Methods in Radio Wave Propagation*; Elsevier: Amsterdam, The Netherlands, 1960; pp. 3–36.
45. Torrent-Moreno, M.; Jiang, D.; Hartenstein, H. Broadcast reception rates and effects of priority access in 802.11-based vehicular ad-hoc networks. In Proceedings of the 1st ACM International Workshop on Vehicular Ad Hoc Networks, Philadelphia, PA, USA, 1 October 2004; pp. 10–18.

46. Issariyakul, T.; Hossain, E. *Introduction to Network Simulator NS2*, 1st ed.; Springer: Berlin/Heidelberg, Germany, 2010.
47. Hanle, C.; Hofmann, M. Performance comparison of reliable multicast protocols using the network simulator ns-2. In Proceedings of the 23rd Annual Conference on Local Computer Networks, Boston, MA, USA, 11–14 October 1998; pp. 222–237.
48. Yinfei, P. *Design Routing Protocol Performance Comparision in NS2: AODV Comparing to DSR as Example*; Department of Computer Science, SUNY Binghamton: Binghamton, NY, USA, 2010.
49. Johnson, D.B.; Broch, J.; Hu, Y.C.; Jetcheva, J.; Maltz, D.A. The cmu monarch project's wireless and mobility extensions to ns. In Proceedings of the 42nd Internet Engineering Task Force, Chicago, IL, USA, 23–28 August 1998; p. 58.
50. Chen, Q.; Schmidt-Eisenlohr, F.; Jiang, D.; Torrent-Moreno, M.; Delgrossi, L.; Hartenstein, H. Overhaul of IEEE 802.11 modeling and simulation in ns-2. In Proceedings of the 10th ACM Symposium on Modeling, Analysis, and Simulation of Wireless and Mobile Systems, Crete Island, Greece, 22–26 October 2007; pp. 159–168.
51. Henderson, T.R.; Roy, S.; Floyd, S.; Riley, G.F. ns-3 project goals. In Proceedings of the 2006 Workshop on ns-2: The IP Network Simulator, Pisa, Italy, 10 October 2006; p. 13.
52. Saeed, T.; Gill, H.; Fei, Q.; Zhang, Z.; Loo, B.T. *An Open-Source and Declarative Approach towards Teaching Large-Scale Networked Systems Programming*; ACM SIGCOMM Education Workshop, Toronto, ON, Canada, 2011.
53. Arbabi, H.; Weigle, M.C. Highway mobility and vehicular ad-hoc networks in ns-3. In Proceedings of the Winter Simulation Conference, Baltimore, MD, USA, 5–8 December 2010; pp. 2991–3003.
54. Treiber, M.; Hennecke, A.; Helbing, D. Congested traffic states in empirical observations and microscopic simulations. *Phys. Rev. E* **2000**, *62*, 1805. [CrossRef]
55. Treiber, M.; Kesting, A. Modeling lane-changing decisions with MOBIL. In *Traffic and Granular Flow'07*; Springer: Berlin/Heidelberg, Germany, 2009; pp. 211–221.
56. Dupont, B. Improvements in VANET Simulator in ns-3. Master's Thesis, Department of Computer Science, Old Dominion University, Norfolk, VA, USA, 2011.
57. Behrisch, M.; Bieker, L.; Erdmann, J.; Krajzewicz, D. SUMO—Simulation of Urban MObility: An overview. In Proceedings of the SIMUL 2011 the Third International Conference on Advances in System Simulation, Barcelona, Spain, 23–29 October 2011; pp. 63–68.
58. Sommer, C.; German, R.; Dressler, F. Bidirectionally coupled network and road traffic simulation for improved IVC analysis. *IEEE Trans. Mob. Comput.* **2011**, *10*, 3–15. [CrossRef]
59. Bilalb, S.M.; Othmana, M.; ur Rehman Khana, A. A performance comparison of network simulators for wireless networks. *arXiv* **2013**, arXiv:1307.4129.
60. Martinez, F.J.; Toh, C.K.; Cano, J.C.; Calafate, C.T.; Manzoni, P. A survey and comparative study of simulators for vehicular ad hoc networks (VANETs). *Wirel. Commun. Mob. Comput.* **2011**, *11*, 813–828. [CrossRef]
61. Khan, A.R.; Bilal, S.M.; Othman, M. A performance comparison of open source network simulators for wireless networks. In Proceedings of the 2012 IEEE International Conference on Control System, Computing and Engineering, Penang, Malaysia, 23–25 November 2012; pp. 34–38.
62. Mittal, N.M.; Choudhary, S. Comparative study of simulators for vehicular ad-hoc networks (vanets). *Int. J. Emerg. Technol. Adv. Eng.* **2014**, *4*, 528–537.
63. Pinart, C.; Sanz, P.; Lequerica, I.; García, D.; Barona, I.; Sánchez-Aparisi, D. DRIVE: A reconfigurable testbed for advanced vehicular services and communications. In Proceedings of the 4th International Conference on Testbeds and research infrastructures for the development of networks & communities, Innsbruck, Austria, 18–20 March 2008; p. 16.
64. Secinti, G.; Canberk, B.; Duong, T.Q.; Shu, L. Software defined architecture for VANET: A testbed implementation with wireless access management. *IEEE Commun. Mag.* **2017**, *55*, 135–141. [CrossRef]
65. Paranthaman, V.V.; Ghosh, A.; Mapp, G.; Iniovosa, V.; Shah, P.; Nguyen, H.X.; Gemikonakli, O.; Rahman, S. Building a prototype vanet testbed to explore communication dynamics in highly mobile environments. In Proceedings of the International Conference on Testbeds and Research Infrastructures, Hangzhou, China, 14–15 June 2016; pp. 81–90.
66. Ahmad, S.A.; Hajisami, A.; Krishnan, H.; Ahmed-Zaid, F.; Moradi-Pari, E. V2V System Congestion Control Validation and Performance. *IEEE Trans. Veh. Technol.* **2019**. [CrossRef]

67. SAE J2945/1_201603. On-Board System Requirements for V2V Safety Communications. Society of Automotive Engineers. 2016. Available online: https://www.its.dot.gov/research_archives/connected_vehicle/pdf/J2945_1_TSS_TP_Test_Specification-20160405.pdf (accessed on 10 May 2019).
68. Heidemann, J.; Bulusu, N.; Elson, J.; Intanagonwiwat, C.; Lan, K.C.; Xu, Y.; Ye, W.; Estrin, D.; Govindan, R. Effects of detail in wireless network simulation. In Proceedings of the SCS Multiconference on Distributed Simulation, Phoenix, AZ, USA, 3–11 January 2001, pp. 3–11.
69. Cavin, D.; Sasson, Y.; Schiper, A. On the accuracy of MANET simulators. In Proceedings of the Second ACM International Workshop on Principles of Mobile Computing, Toulouse, France, 30–31 October 2002; pp. 38–43.
70. Kotz, D.; Newport, C.; Gray, R.S.; Liu, J.; Yuan, Y.; Elliott, C. Experimental evaluation of wireless simulation assumptions. In Proceedings of the 7th ACM International Symposium on Modeling, Analysis and Simulation of Wireless and Mobile Systems, Venice, Italy, 4–6 October 2004; pp. 78–82.
71. Torrent-Moreno, M.; Santi, P.; Hartenstein, H. Fair sharing of bandwidth in VANETs. In Proceedings of the 2nd ACM International Workshop on Vehicular Ad Hoc Networks, Cologne, Germany, 2 September 2005; pp. 49–58.
72. Yang, X.; Liu, L.; Vaidya, N.H.; Zhao, F. A vehicle-to-vehicle communication protocol for cooperative collision warning. In Proceedings of the First Annual International Conference on Mobile and Ubiquitous Systems: Networking and Services, Boston, MA, USA, 22–26 August 2004; pp. 114–123.
73. Kabir, T.; Nurain, N.; Kabir, M.H. Pro-AODV (Proactive AODV): Simple modifications to AODV for proactively minimizing congestion in VANETs. In Proceedings of the 2015 International Conference on Networking Systems and Security (NSysS), Dhaka, Bangladesh, 5–7 January 2015; pp. 1–6.
74. Ruehrup, S.; Fuxjaeger, P.; Smely, D. TCP-like congestion control for broadcast channel access in VANETs. In Proceedings of the 2014 International Conference on Connected Vehicles and Expo (ICCVE), Vienna, Austria, 3–7 November 2014; pp. 427–432.
75. Huang, C.L.; Fallah, Y.P.; Sengupta, R.; Krishnan, H. Adaptive intervehicle communication control for cooperative safety systems. *IEEE Netw.* **2010**, *24*, 6–13. [CrossRef]
76. Shah, S.A.A.; Ahmed, E.; Rodrigues, J.J.; Ali, I.; Noor, R.M. Shapely value perspective on adapting transmit power for periodic vehicular communications. *IEEE Trans. Intell. Transp. Syst.* **2018**, *19*, 977–986. [CrossRef]
77. Gupta, P.; Kumar, P.R. The capacity of wireless networks. *IEEE Trans. Inf. Theory* **2000**, *46*, 388–404. [CrossRef]
78. Crash Avoidance Metrics Partnership, Vehicle Safety Communications Consortium. *Vehicle Safety Communications Project: Task 3 Final Report: Identify Intelligent Vehicle Safety Applications Enabled by DSRC*; National Highway Traffic Safety Administration, US Department of Transportation: Washington, DC, USA, 2005.
79. Weinfield, A.; Kenney, J.B.; Bansal, G. An Adaptive DSRC Message Transmission Rate Control Algorithm. In Proceedings of the 18th ITS World CongressTransCoreITS AmericaERTICO-ITS EuropeITS Asia-Pacific, Orlando, FL, USA, 16–20 October 2011.
80. Mittag, J.; Schmidt-Eisenlohr, F.; Killat, M.; Härri, J.; Hartenstein, H. Analysis and design of effective and low-overhead transmission power control for VANETs. In Proceedings of the Fifth ACM International Workshop on VehiculAr Inter-NETworking, San Francisco, CA, USA, 15 September 2008; pp. 39–48.
81. Willis, J.T.; Jaekel, A.; Saini, I. Decentralized congestion control algorithm for vehicular networks using oscillating transmission power. In Proceedings of the 2017 Wireless Telecommunications Symposium (WTS), Chicago, IL, USA, 26–28 April 2017; pp. 1–5.
82. Joseph, M.; Liu, X.; Jaekel, A. An Adaptive Power Level Control Algorithm for DSRC Congestion Control. In Proceedings of the 8th ACM Symposium on Design and Analysis of Intelligent Vehicular Networks and Applications, Montreal, QC, Canada, 28 October–2 November 2018; pp. 57–62.
83. Kloiber, B.; Härri, J.; Strang, T. Dice the tx power—Improving awareness quality in vanets by random transmit power selection. In Proceedings of the 2012 IEEE Vehicular Networking Conference (VNC), Seoul, Korea, 14–16 November 2012; pp. 56–63.
84. Goudarzi, F.; Asgari, H. Non-Cooperative Beacon Power Control for VANETs. *IEEE Trans. Intell. Transp. Syst.* **2018**, *20*, 777–782. [CrossRef]
85. Sepulcre, M.; Gozalvez, J.; Härri, J.; Hartenstein, H. Contextual Communications Congestion Control for Cooperative Vehicular Networks. *IEEE Trans. Wirel. Commun.* **2011**, *10*, 385–389. [CrossRef]

86. Baldessari, R.; Scanferla, D.; Le, L.; Zhang, W.; Festag, A. Joining forces for vanets: A combined transmit power and rate control algorithm. In Proceedings of the 7th International Workshop on Intelligent Transportation (WIT), Hamburg, Germany, 22–23 March 2010. Available online: http://www.festag-n et.de/doc/2010_WIT_powerratecontrol_vanets.pdf (accessed on 10 May 2019).
87. Tielert, T.; Jiang, D.; Hartenstein, H.; Delgrossi, L. Joint power/rate congestion control optimizing packet reception in vehicle safety communications. In Proceedings of the Tenth ACM International Workshop on Vehicular Inter-Networking, Systems, and Applications, Taipei, Taiwan, 25 June 2013; pp. 51–60.
88. Jacobson, V. Congestion avoidance and control. In Proceedings of the ACM SIGCOMM Computer Communication Review, Stanford, CA, USA, 16–18 August 1988; Volume 18, pp. 314–329.
89. Math, C.B.; Li, H.; De Groot, S.H.; Niemegeers, I. A combined fair decentralized message-rate and data-rate congestion control for V2V communication. In Proceedings of the 2017 IEEE Vehicular Networking Conference (VNC), Torino, Italy, 27–29 November 2017; pp. 271–278.
90. Huang, C.L.; Sengupta, R. Decentralized error-dependent transmission control for model-based estimation over a multi-access network. In Proceedings of the 4th Annual International Conference on Wireless Internet, WICON 2008, Maui, HI, USA, 17–19 November 2008; p. 80.
91. Harsanyi, J.C.; Selten, R. *A General Theory of Equilibrium Selection in Games*; MIT Press Books: Cambridge, MA, USA, 1988; Volume 1.
92. He, J.; Chen, H.H.; Chen, T.M.; Cheng, W. Adaptive congestion control for DSRC vehicle networks. *IEEE Commun. Lett.* **2010**, *14*, 127–129. [CrossRef]
93. Lan, T.; Kao, D.; Chiang, M.; Sabharwal, A. *An Axiomatic Theory of Fairness in Network Resource Allocation*; IEEE: Piscataway, NJ, USA, 2010.
94. Sepulcre, M.; Gozalvez, J.; Altintas, O.; Kremo, H. Integration of congestion and awareness control in vehicular networks. *Ad Hoc Netw.* **2016**, *37*, 29–43. [CrossRef]
95. Sepulcre, M.; Gozalvez, J. Coordination of Congestion and Awareness Control in Vehicular Networks. *Electronics* **2018**, *7*, 335. [CrossRef]

© 2019 by the authors. Licensee MDPI, Basel, Switzerland. This article is an open access article distributed under the terms and conditions of the Creative Commons Attribution (CC BY) license (http://creativecommons.org/licenses/by/4.0/).

Review

A Survey on Fault Tolerance Techniques for Wireless Vehicular Networks

João Almeida [1,*], João Rufino [1], Muhammad Alam [1] and Joaquim Ferreira [1,2]

1. Instituto de Telecomunicações, Campus Universitário de Santiago, 3810-193 Aveiro, Portugal; joao.rufino@ua.pt (J.R.); alam@ua.pt (M.A.); jjcf@ua.pt (J.F.)
2. ESTGA—Universidade de Aveiro, 3754-909 Águeda, Portugal
* Correspondence: jmpa@ua.pt

Received: 7 October 2019; Accepted: 12 November 2019; Published: 16 November 2019

Abstract: Future intelligent transportation systems (ITS) hold the promise of supporting the operation of safety-critical applications, such as cooperative self-driving cars. For that purpose, the communications among vehicles and with the road-side infrastructure will need to fulfil the strict real-time guarantees and challenging dependability requirements. These safety requisites are particularly important in wireless vehicular networks, where road traffic presents several threats to human life. This paper presents a systematic survey on fault tolerance techniques in the area of vehicular communications. The work provides a literature review of publications in journals and conferences proceedings, available through a set of different search databases (IEEE Xplore, Web of Science, Scopus and ScienceDirect). A systematic method, based on the preferred reporting items for systematic reviews and meta-analyses (PRISMA) Statement was conducted in order to identify the relevant papers for this survey. After that, the selected articles were analysed and categorised according to the type of redundancy, corresponding to three main groups (temporal, spatial and information redundancy). Finally, a comparison of the core features among the different solutions is presented, together with a brief discussion regarding the main drawbacks of the existing solutions, as well as the necessary steps to provide an integrated fault-tolerant approach to the future vehicular communications systems.

Keywords: wireless vehicular communications; systematic review; fault tolerance; dependability

1. Introduction

The main motivation behind the development of wireless vehicular communications was based on the need to improve road safety and traffic efficiency. Following previous success cases in the field of intelligent transportation systems (ITS), vehicular networks hold the potential of drastically reducing the number of traffic accidents and road fatalities. For that purpose, international standards have been purposed with the goal of defining how safety messages should be exchanged among vehicles and between vehicles and the roadside infrastructure. IEEE WAVE and ETSI ITS-G5 constitute the most disseminated protocol stacks for vehicular communications in the United States and Europe, respectively. Both of them rely on the IEEE 802.11 standard for the implementation of physical and medium access control (MAC) layers. Recent proposals, such as LTE-V or 5G, are pointed to as possible alternatives to 802.11, since these allow several distinct applications (e.g., mobile broadband, railway systems, etc.) to share the cost of network installation.

Despite the focus on safety, vehicular communications systems by design typically do not take into consideration dependability attributes and the need for hard real-time guarantees. However, this type of safety-critical services demands small end-to-end delays and high levels of reliability and availability. Fault tolerance mechanisms are good candidates to attain such requirements [1]. In this survey, a review of fault tolerance techniques for vehicular networks is presented. A systematic method was followed in order to select the related articles from a set of different search databases.

Fault tolerant methods are sometimes employed in security mechanisms for safety-critical applications in vehicular networks. In fact, the concepts of dependability and security are closely related [2] and in some cases, the proposed solutions aim to increase both the dependability and security attributes of the network. Nevertheless, an effort was made on this survey to avoid the analysis of strategies focusing on the security issues of vehicular communications, since this could be the topic of another review paper, given the large number of threats already identified in the literature [3]. In addition, the search was limited to fault-tolerant techniques specifically addressing vehicular network issues, not broader research topics such as mobile ad-hoc networks (MANETs), or similar fields e.g., mobile sensor networks (MSNs). An example of related work done in fault-tolerant MSNs, which usually take into consideration energy constraints that are not present in vehicular systems, is the mobility-tolerant TDMA-based MAC protocol proposed by Jhumka and Kulkarni [4].

In real-time wireless communication systems such as vehicular communication for safety applications, high availability of the system is very important to guarantee the dissemination of the safely critical messages to the desired destinations in the bounded time. Fault tolerance prevents the connection disruption arising from the system's component failures and therefore, high availability is achieved by ensuring no loss of service. Since fault tolerant systems provide real-time backup and usually depend on the redundant components, they are associated with additional costs. In addition, fault tolerant techniques prevent the failures and therefore restrict the scope of these failures in distributed systems.

The rest of the paper is organized as follows. Section 2 provides some background regarding the topic of fault tolerant communications and the different types of redundancy techniques. Section 3 presents the search method and the paper selection process for this survey based on the discussed criteria. In Section 4, the selected papers are analysed and classified according to the type of redundancy technique used to provide fault-tolerant behaviour. A comparison among some core characteristics of the selected documents is also provided, together with a discussion regarding the relevant findings of this survey. Finally, Section 5 presents the main conclusions of this work.

2. Background

Fault tolerant communications aim to guarantee that two or more network nodes can exchange information in spite of faults that may affect the communications link or some of the participating nodes. A communication system providing fault tolerance capabilities typically involves redundancy and diversity techniques. This way, it is possible to avoid the presence of single points of failure in the system and to prevent common failure modes in network nodes or node's components. Communications redundancy corresponds to an increment in resource utilization (mainly replication), in order to provide resilience against faults arising in the system. Traditionally, redundancy techniques can be classified into three main groups:

- **Temporal redundancy** is characterized by the attempt to deliver the same information at multiple moments in time. Retransmission based protocols, such as TCP, are clear examples of this strategy.
- **Information redundancy** corresponds to the use of additional data, so that information can still be retrieved in case of partial data loss. For instance, error correction codes require the transmission of redundant data in order to recover the contents of the exchanged message.
- **Spatial redundancy** refers to the possibility of providing the same information from different sources. Hardware replication constitutes a traditional example of spatial redundancy, where several replicas are able to deliver the same service.

Redundancy may also be categorized in terms of protocol stack layer in which it is applied. For instance, redundant channel links can be classified as physical layer redundancy. As a result, the distinct strategies may be categorized based on the Open Systems Interconnection (OSI) model. Cross-layer solutions are also possible, covering faults in a set of different layers of the protocol stack. For example, entire node replication, from the RF antenna to the application level, constitutes one of these cases.

Fault-tolerant wireless communications strategies can also be found in other areas of research beyond vehicular networks. For instance, in wireless sensor networks (WSNs) or other types of mobile ad-hoc networks (MANETs), there are also available solutions in the literature to deal with possible faults in the system's operation. However, vehicular communications systems pose specific challenges, ranging from very dynamic network topologies to frequent link disruptions and the Doppler effect. As a result, for this survey, only fault-tolerance techniques particularly targeting vehicular network applications were considered.

3. Method

3.1. Search Approach and Groups of Keywords

The search method for this survey was based on a systematic review process according to the preferred reporting items for systematic reviews and meta-analyses (PRISMA) statement [5]. Four different databases were utilized in this search (IEEE Xplore, Web of Science, Scopus and Science Direct), while others, such as Google Scholar, TRID or Academic Search Complete, were also explored but due to several distinct restrictions (e.g., the impossibility to search only on the article's metadata), were not included in the search tools for this review. The approach followed in the paper identification process, required that at least one term from each of two groups of keywords was present in the article's metadata. These two groups of keywords were the following:

- "fault tolerance", "fault tolerant", "fault detection", "dependability", "dependable", "safety critical"
- vehicular network*, vehicular communication*, "connected vehicles", "VANET", "cooperative vehicles", "cooperating vehicles", "intervehicle communications", "vehicle to vehicle"

The terms "reliability", "reliable" and "real-time" were also considered for the first group of keywords, however, since these terms are typically employed in a broader sense, it was decided to limit the search to the keywords shown above. By employing these search terms, a total number of 1493 papers were identified (586 from the IEEE Xplore database, 299 from Web of Science, 585 from Scopus and 23 from ScienceDirect).

Both Scopus and Web of Science search tools provide charts with the results distribution along the years. These graphs can be observed in Figure 1. From the analysis of both charts, it is possible to derive a clear growth in the number of articles published with the search terms. This trend demonstrates the raising importance given by the scientific community to the topic of fault tolerance and dependability in the field of vehicular networks.

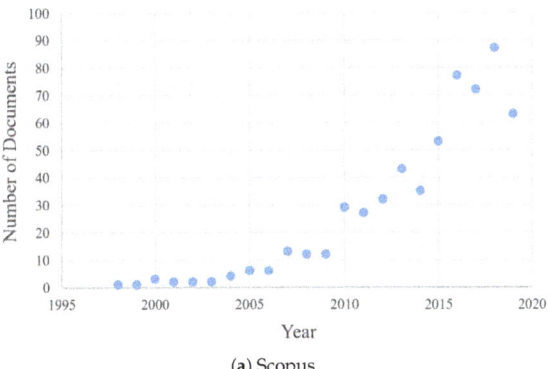

(a) Scopus.

Figure 1. Cont.

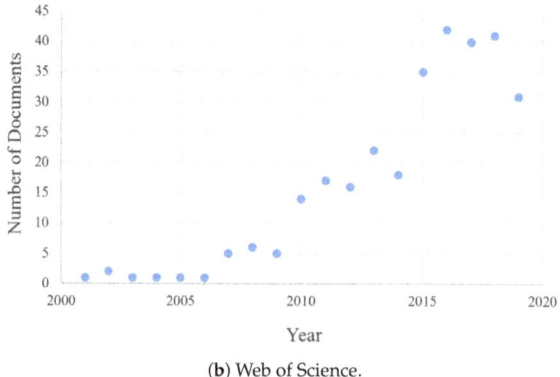

(b) Web of Science.

Figure 1. Results distribution along the years (adapted from Scopus and Web of Science databases).

3.2. Selection Process and Inclusion/Exclusion Criteria

After this initial step, the duplicated entries were removed, as well as standards and other invalid results, such as programs, forewords or table of contents from conference proceedings. A total of 819 papers remained for analysis. Then, the titles and abstracts of the remaining articles were screened to exclude research work outside the scope of this survey, focusing for instance on satellite communications or intra-vehicle networks such as Controller Area Network (CAN) or FlexRay. 172 records were still left for the final selection process, in which not only the metadata (title and abstract) but also the body of the paper was analysed. A set of several criteria was used to ensure the eligibility of the articles to include in this survey:

- be published in English in a peer reviewed journal or conference proceedings;
- be focused on safety applications using wireless vehicular communications;
- including fault tolerance techniques to improve the dependability attributes of vehicular networks.

Complementarily, the following exclusion criteria were utilized in this selection process:

- only the most recent paper from a set of similar articles by the same author was kept (for instance, a paper published in a conference and later extended to a journal or magazine);
- papers not specifically focusing on safety-critical applications (with strict real-time constraints) were excluded;
- articles focusing on security issues of vehicular networks were not selected for the review;
- papers targeting railway, aviation systems, unmanned autonomous vehicle (AUxV), military vehicular clouds (MVC) or wireless sensor networks (WSN) were also considered out of the scope of this analysis;
- research works referring to fault-tolerance techniques designed to alternative technologies for vehicular applications, such as visible light communications (VLC), were not included;
- articles not specifically focusing on vehicular networks were discarded, like the ones with a broader scope (e.g., MANETs).

Following these criteria, 24 papers were selected. An additional record was added after screening the reference lists of this final selection, summing up a total number of 25 articles to be the subject of analysis in this survey. The complete paper selection process can be visualized in Figure 2.

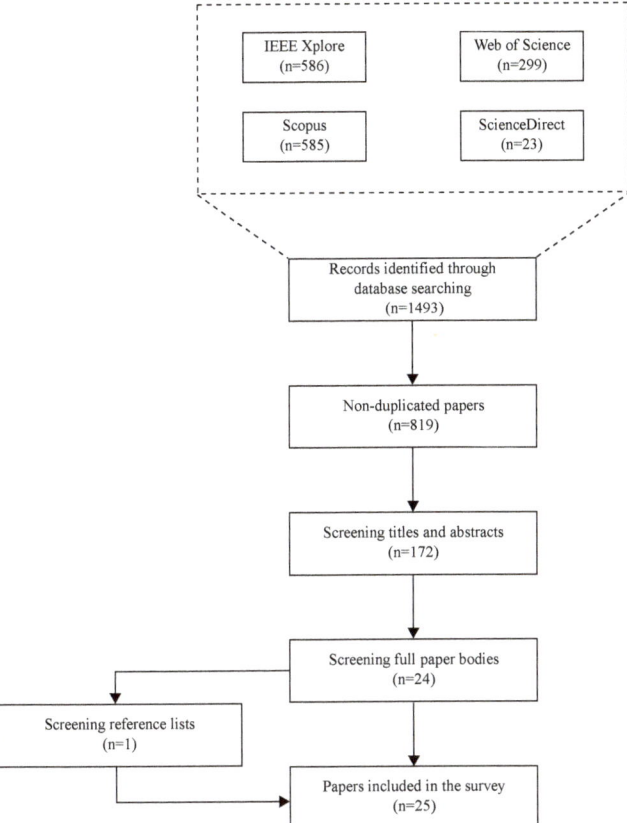

Figure 2. Systematic paper review process according to preferred reporting items for systematic reviews and meta-analyses (PRISMA) statement [5].

4. Fault-Tolerance Techniques

The selected publications encompass fault tolerance techniques that can be categorized according to the type of communications redundancy used (temporal, spatial and information). This classification can be visualized in Table 1. This section summarizes the contributions of each one of the selected publications.

Table 1. Results classification according to redundancy types.

Temporal	Spatial	Information
Jonsson et al. [6]	Matthiesen et al. [7]	Kumar et al. [8]
Bohm et al. [9]	Cambruzzi et al. [10]	Eze et al. [11]
Savic et al. [12]	Abrougui et al. [13]	Ali et al. [14]
Sawade et al. [15]	Chang et al. [16]	
Nguyen et al. [17]	Lann et al. [18]	
	Sanderson et al. [19]	
	Aljeri et al. [20]	
	Casimiro et al. [21]	
	Worral et al. [22]	
	Ploeg et al. [23]	

Table 1. *Cont.*

Temporal	Spatial	Information
	Fathollahnejad et al. [24]	
	Bhoi et al. [25]	
	Elhadef et al. [26]	
	Almeida et al. [27]	
	Medani et al. [28]	
	Devangavi et al. [29]	
	Younes et al. [30]	

4.1. Temporal Redundancy

Regarding temporal redundancy methods, solutions found in the literature are based on packet retransmission schemes. For instance, the work of Jonsson et al. [6] focuses on increasing MAC protocol reliability for platooning applications. Time is divided into periodic transmission cycles, called superframes, each one including both contention-based phases (CBPs) and contention-free phases (CFPs). Applications with hard real-time constraints utilize the CFPs to transmit information. These CFPs rely on a polling-based mechanism administered by a master vehicle that applies a time division multiple access (TDMA) scheme for nodes' transmissions. Each transmission corresponds to a specific time slot, which are ordered according to the earliest deadline first (EDF) policy and a real-time automatic repeat request (ARQ) scheme. This ARQ scheme allows the retransmission of packets to not be well received and they can still be transmitted before their deadline expires. The performance evaluation of the protocol demonstrates a reduction in message error rate by several orders of magnitude when compared to the case without retransmissions.

Despite the improved reliability provided by the proposed solution, this retransmission scheme requires some modifications to the transport layer of vehicular communications protocol stack, in order to be implemented. This non-compliance with the standards encompasses the addition of a real-time polling-based layer on top IEEE 802.11p MAC, as well as the implementation of the transport layer retransmission scheme over a service channel exclusively dedicated to platooning communications. Another disadvantage is the fact that in highly congested scenarios, the retransmission scheme can lead to some bandwidth reduction for each transmitter, however, the real-time scheduling algorithm (EDF) guarantees the optimization of channel use. Furthermore, nothing is referred about the possibility of master failure, the one responsible for coordinating all the communications in the platoon, and how the proposed scheme reacts to that event. Finally, it should be also pointed out that the evaluation of this work was only performed in a simulation environment, very similarly to a numerical analysis, in which the network parameters were kept very static.

Following a similar approach, Böhm and Kunert [9] propose a retransmission scheme based on the data age of previously received messages. The framework targets intra-platooning communications but also communication between different platoons (inter-platooning). A dedicated service channel is used for intra-platooning communications, while vehicles in distinct platoons exchange information through the control channel. The platoon leader or master is responsible for periodically disseminating beacon messages to all the other vehicles in the platoon. Then during a collection phase, vehicles transmit status updates, which may or may not be well received by the master. In case of unsuccessfully decoded packets, the leader vehicle initiates a retransmission phase by sending individual polling messages to the other nodes, that immediately attempt to retransmit the failed messages. After that, a control phase begins which is used by the master to coordinate the other platoon members based on the retrieved status information. During this window, control packets are transmitted individually to each regular vehicle. In the end, another retransmission phase begins based on acknowledgements returned by the receiving nodes. Moreover, retransmission opportunities are assigned to the nodes, according to the data age of the messages received by the leader vehicle, which keeps a table with the reception time of the latest status update and acknowledgement frames. From this record, higher transmission priorities are allocated to vehicles with older successfully transmitted messages.

The proposed solution introduces some tweaks at the MAC and transport layers in relation to the standard protocols and requires a significant amount of overhead, with acknowledgment messages, retransmissions packets and individual polling messages, in order to improve the packet delivery ratio, making the available bandwidth smaller for other applications or in case of channel saturation. Moreover, the authors do not take into account the problem of having the platoon coordinator as a single point of failure, holding the table with data age of messages from each vehicle. The protocol evaluation demonstrated the feasibility of the proposed scheme and the ability to maintain a stable data age value for the platoons. Additionally, the simulation analysis compares the proposed protocol with the standard approach and other retransmission schemes, but lacks a close-to-real-world scenario evaluation.

The work of Savic et al. [12] targets a distinct application, in this case, the collision avoidance problem of fully autonomous cars at road intersections. An algorithm for distributed intersection crossing is proposed, being able to cope with an unknown and large number of communications failures. A priority for intersection crossing is assigned to each one of the vehicles based on current position estimates and the cars' dynamics. Three types of packets are exchanged: periodic heartbeat messages and 'ENTER' and 'EXIT' messages immediately before and after crossing the intersection, respectively. In case of receive-omission failures, the 'ENTER' or 'EXIT' packets are retransmitted and the model assumes that at least one heartbeat ('HB') message is received before the intersection crossing (IC) algorithm starts and that vehicles eventually succeed to receive the 'ENTER' and 'EXIT' messages. The numerical results show only a slight increase of the crossing delay in the presence of communications failures.

The limitations of the proposed model include the omission of transmitter faults, messages with erroneous content and the assumptions of receiving at least one 'HB' packet prior to the initialization of the IC algorithm and the eventual successful reception of the sent 'ENTER' and 'EXIT' messages. An advantage of this solution is the fact that there is no centralized entity to control the intersection crossing, avoiding single point of failures in the system. Regarding the evaluation process, the numerical analysis is very limited since it only considers two vehicles and consecutive receive-omission failures. Further testing with a high number of vehicles and with more arbitrary conditions in common traffic simulators and real-world implementations is necessary, in order to better evaluate the proposed algorithm.

Sawade et al. [15] propose a protocol for cooperative maneuvers under adverse conditions. The proposed solution relies on bidirectional stateful communication, i.e., an established session link connecting two or more participants on a collaborative driving maneuver. A synchronization layer is added on top of the bi-directional negotiation of collaborative maneuvers, through the utilization of the Turquois algorithm for attaining distributed consensus under byzantine conditions. Participants must send heartbeat periodic messages in order to keep the session open. Once a predefined number of consecutive missed messages from a vehicle, the session is called unstable and can be terminated. Once in a session, any vehicle broadcasts the current session state in a hashed value. The session state must be consensual across the party of vehicles, so each bit of the hash is individually synchronized between stations. A parameter is used to control the robustness of the sessions against packet loss. This factor is a tradeoff between packets lost consecutively before a failure state is requested and the assurance of consensus among vehicles.

This work adds missing capabilities to the existing vehicular communications standards, through the integration of a collaborative maneuver protocol in the ETSI ITS-G5 protocol stack. The proposal has the advantage of introducing new features on the message-layer only, thus being backwards-compatible to current standard implementations. The main drawback of the work, however, is on the evaluation part, since only simulation results are provided for the specific case of a platoon with just two vehicles. Nevertheless, the conducted experiments show that a session would be stable 99% of the time for reasonable tradeoff values and environments with less than 20% of packet loss.

Nguyen et al. [17] also propose a protocol that encompasses the retransmission of safety messages that failed to broadcast. The protocol takes into account the presence of hidden nodes and their effect on communications faults. The proposed multi-channel MAC scheme (RAM) divides the control channel cycle in three main intervals: the safety interval, the response interval and the contention-based interval. Any collided safety packets can be retransmitted in the contention-based interval. Whenever a vehicle does not receive any safety packet within a time window from its neighbours, it will request an RSU to send one. The RSUs behave as a central authority inside a given area, being responsible for managing the duration of each cycle, tracking the exchanged messages and adverting the vehicles what packets were successfully received. Based on vehicle density and data traffic conditions, the RSU optimizes the length of the contention-based interval, by also taking into consideration the hidden terminal problem. A Markov chain model is used to analyze the reliability of the real-time transmission of safety packets and to provide information for the computation of the optimized control channel intervals. Simulation results show the improved performance of the proposed RAM protocol in terms of packet delivery ratio in comparison with two other related works.

Despite the increased reliability in the transmission of safety packets, the proposed RAM scheme cannot be directly applied using current vehicular communications equipment, since it requires some modifications to the standard MAC layer, due to the division of the control channel interval into three distinct phases: one congestion-based period (as the standard MAC protocol operates) and two congestion-free intervals. The protocol also introduces some additional overhead in the communications, as a result of the need to transmit acknowledgment messages and retransmission packets. Moreover, the solution has the drawback of assuming that vehicles are distributed along a straight line, in order to simplify the hidden terminal problem, which is typically not the case in real world conditions, where several roads interconnect with a lot of physical obstacles in the middle, either in urban or highway environments. The simulation results lack the implementation in standard traffic and network simulators software and the diversity of simulated traffic environments.

4.2. Spatial Redundancy

In [7], Matthiesen et al. investigate the utilization of replicated application services in dynamic clusters of vehicles. The goal is to increase the reliability and availability of safety-critical applications in ad-hoc vehicular networks. The example of a distributed shared memory, which supports the operation of a stateful road-traffic information service, is presented in this work. Several metrics are analysed and evaluated for different cluster dimensions, such as data consistency, response time and application availability. A Replication Manager is employed in order to achieve stable clusters, by selecting replicas with good communication metrics that minimize service response time and reconfiguration overhead in case of faulty behaviour. These faults can be due for instance to excessive delay or high packet loss, which may affect timeliness and correctness of the service, thus leading to inconsistent application states.

The proposed fault-tolerant model has the advantage of not requiring any changes in the protocol stack, since the replication model is fully deployed at the application layer. On the other hand, however, overhead of replica selection and exchanging servers in case of failure is not taken into account and may have a significant impact in real-world operation, due to the topology changes and very dynamic environments in which vehicular networks operate. The model also does not consider network congestion scenarios, where the proposed solution may not operate as expected. With simulation results or real test-case measurements, these last points could be better evaluated, not being limited to the the numerical analysis provided to validate only some parameters of the replication service.

The work of Cambruzzi et al. [10] proposes a failure detection scheme based on a protocol that detects both link and system failures. It employs a heartbeat mechanism in which all roadside units and vehicles transmit a beacon message periodically to their single-hop neighbours. When a beacon packet reaches its destination, the receiving node adds or updates its neighbours' list with the received information together with a timestamp of the packet. If no message is received from that neighbour during a predefined time interval, the node is considered to be faulty and is inserted into a list of suspects. The algorithm uses adaptive timeouts in order to cope with the dynamic conditions of vehicular communication networks. In this model, only two types of faults are assumed. Those caused by a system crash and the ones caused by a vehicle exiting the road. Malicious or Byzantine faults are not considered in this study.

The fault model considered in the design of this failure detection scheme is very limited, since it only takes into account two type of faults, namely crash-faults (in case of equipment crash) and exit-faults (when a vehicle exits the road). For instance, babbling idiot failures are not analysed, which restricts the validity of the proposed model. Moreover, the impact of the exchange of tables, with the list of neighbours and their perceived status, in the communications overhead is neither discussed nor analysed. Finally, the simulation experiments consider only a simplified scenario with a straight road segment where all vehicles move in the same direction. In practice, this scenario may only happen in very few cases and, therefore, more complex environments should be evaluated, since the model depends significantly on the variation of network topology and link stability.

Based on a similar failure detection mechanism, Abrougui et al. [13] introduce a fault-tolerance location-based service discovery protocol for vehicular networks. This protocol handles the discovery procedure of different types of both safety and infotainment services and it was designed to perform well even in the presence of service provider failures, communication link failures and roadside units failures. The proposal relies on a cluster-based infrastructure, where roadside units are clustered around service providers, the congested areas of the vehicular network and the intermittent areas to improve the connectivity of the network. The proposed fault-tolerance mechanisms were introduced at the network level, in order to cope with several types of failures in the connection between the service provider and the service requester. Essentially, in case of link or system failure, an algorithm is employed to designate alternative nodes that will supply or forward the information missed in the faulty nodes/links. Simulation results showed an improvement in the success rate of discovery queries of approximately 50% and 30%, in case of a roadside router and link failure scenarios, respectively, when compared with a simplified version of the protocol without fault tolerance techniques.

However, this fault-tolerance scheme presents some disadvantages, such as the fact that the routing protocol is based on a non-standard solution (CLA-S), which requires some modifications in the protocol stack. Additionally, it introduces overhead in the communications protocol, by requiring mechanisms such as the leader election for the roadside routers. Despite the fact that the proposed fault detection mechanism is also able to detect intermittent failures, only permanent ones are considered in the simulation experiments. Moreover, no measurements of the time to recover from failures, e.g., including failure detection time and leader reelection phase, are presented.

Chang and Wang [16] propose a fault-tolerant protocol for a reliable broadcast of alert messages in vehicular ad-hoc networks. The goal of this protocol is to reduce the total number of messages needed to disseminate the alert message along the road. The proposed method designates the two farthest vehicles in the radio range of the source vehicle to act as candidate relay nodes of the message to be broadcast. This selection is performed by the source vehicle and it is based on the GPS coordinates provided by all vehicles in the transmission range. If the farthest vehicle from the source node does not transmit the safety message within a maximum time interval, the sub-farthest will assume that there was a system failure and will disseminate the intended message. The results show that the penetration rate of the fault-tolerant scheme is very satisfactory even for low traffic densities, providing advantages in relation to the simple flooding method in terms of transmission delay and total number of messages exchanged in the wireless medium.

The proposed protocol has the limitation of being specifically designed for network topologies typically found in the highway scenarios. For instance, in urban environments with a lot of road intersections, it could be important to disseminate warning messages in different directions. In such context, this solution with only one farthest and one sub-farthest vehicles can no longer be applied. In addition to this, the protocol requires some modifications both to the standard MAC and transport layers, which means that it cannot be directly deployed using commercial off-the-shelf (COTS) components. Furthermore, the simulation experiments could include not only the testing of natural communications link limitations, but also communications faults and equipment crashes, in order to broaden the scope of the fault model analysis.

The work of Gérard Le Lann [18] deals with omission failures originated by a transient fault in the transmitter, receiver or in the communications channel. High reliability and strict timeliness properties are achieved through group dissemination protocols so that every message can be delivered to a given set of vehicles within a worst-case deadline. A Zebra protocol suite which comprises geocast, convergecast, multicast and the Altruistic protocol is employed to guarantee the timely delivery of messages. The proposed fault-tolerant strategy relies on the spatial redundancy provided by the multiple copies of information kept in the different vehicles. This approach would typically lead to high overhead, however, the notion of proxy sets is introduced in order to limit the scope size of the global dissemination protocol.

The proposed scheme has the main drawbacks of not considering permanent failures, i.e., equipment crashes, but only transient faults, and the fact that it specifically targets platooning applications, not being designed as a more generic solution for other safety-critical vehicular applications. It also requires changes to the standard protocol stack, namely at the MAC, routing and transport layers, by employing a suite of protocols (Zebra), specifically designed for time-critical single hop multipoint communications. Besides, it needs a more thoughtful evaluation, since neither simulation nor real test-case experiments were conducted.

Sanderson and Pitt [19] propose an adaptation of the *Paxos* algorithm [31] to implement consensus formation in self-organizing vehicular networks. The proposed algorithm (*IPcon*) handles institutionalized consensus in spite of faults occurring in the dynamic clusters of vehicles. The protocol tolerates faults caused by nodes that fail by permanently stopping or later restarting, delayed, lost or duplicated messages, however, malicious vehicles and corrupted packets are not considered. The evaluation of the algorithm demonstrates the resilience against role failures (nodes may play four different roles in the *IPcon* protocol) and cluster fragmentation and aggregation.

One of the limitations of the proposed solution is that it does not take into account all faults in the value domain, e.g., corrupted message content. Moreover, the communications overhead of the consensus algorithm (IPCon) may have some negative impact on the timeliness of safety-critical applications running on top of vehicular networks. Similarly, the leader election and conflict resolution processes could also consume a considerable amount of time to be executed. Practical evaluation regarding these time measurements should be carried out, in order to assess the validity of the proposed scheme under dynamic real-world scenarios.

A fault detection protocol is introduced in [20] by Aljeri et al. in order to mitigate communications problems in vehicular networks. Fault diagnosis is performed by comparing the output messages from a group of vehicles. This way, it is possible to identify faulty vehicular nodes. The process is initiated by an RSU, which attributes the same task to a group of vehicles. Then, the results are computed by each node and the answers are transmitted back to the initiator. If the results are identical, it is assumed that there are no faults in the network. On the other hand, if different results are yielded, faulty road components can be detected. Additionally, a more efficient implementation of the protocol is proposed that relies on regional RSUs, which decreases the total number of packets transmitted and the diagnosis latency of this method.

The proposed fault detection mechanism implies additional network resource usage, in order to identify faulty nodes. The tasks assigned to pairs of vehicles, with the goal of verifying disagreements and diagnosing faults, introduce some communications rounds and consume processing time. It is not a transparent solution that takes advantage of the messages already exchanged inside the vehicular network. Additionally, it is assumed that two faulty vehicles always give different outputs, which may not always be the case, e.g., in common failure mode. Another drawback relies on the fact that the fault detection scheme only targets networks where roadside units are present. There is no alternative framework devised for communications solely among clusters of vehicles, or for the specific case when there is a permanent failure in the gateway node, which behaves as a single point of failure in the network.

In [21], Casimiro et al. develop a kernel-based architecture (KARYON) for safety-critical coordination in vehicular systems. Besides dealing with sensor faults and real-time properties of the wireless communications (e.g., self-stabilizing protocol), the proposed architecture also introduces extra components to the standard MAC layer. According to the followed subsystem isolation, the authors assume in the fault model that communication components can experience crash or timing faults, however, data cannot be corrupted, i.e., faults may occur in the time domain, but not in the value one. In addition to the standard MAC layer, two extra elements are introduced: the mediator Layer (MLA) and the Channel Control Layer. For example, the MLA is responsible for node failure detection and membership and control of temporary network partitions. On the other hand, the channel control layer supervises the channel state and enhances the network resilience by taking advantage of the diversity of radio channels available for vehicular communications purposes.

The main disadvantage of the devised architecture is the need for introducing several changes in the protocol stack, especially at the MAC layer level, so that some extra functionalities that are not present in current COTS components become available. As already mentioned, in the communications modules of the system, not all types of faults are covered, since crash or timing faults are tolerated, but not data corrupted messages. Finally, no evaluation is performed in this publication, that is part of the future work, so there is no way to validate the performance of the proposed fault-tolerant solution.

The work of Worrall et al. [22] deals not only with the complete loss of radio communications but also with partial degradation of the wireless link. In some cases, the communications performance is affected in an intermittent way or behaves poorly after a certain distance, due e.g., to damages in the external cables, antennas or connector. The proposed method utilizes data gathered during normal operation so that the antenna behaviour can be modelled and used in future fault detection. This model is derived by analysing and learning the properties of wireless communications in a fleet of vehicles, taking into account parameters such as relative orientation, bearing and range between vehicles. The detailed knowledge about the communications performance is then utilized to detect partial antennas faults or permanent link failures, which are identified by observing when the RF communications deviate from the expected operation. Additional computational resources are required in order to allow online execution of the mathematical model and appropriate comparison with the run-time results of the antenna performance.

The proposed fault detection mechanism is not suitable for a large number of vehicular communications applications, which are based on broadcast messages, since this solution is specifically designed for point-to-point radio links. The scheme only covers faults in the physical air interface, namely cable and antenna performance degradation, not detecting any time and value issues in the exchanged messages. Furthermore, the real test case results show that the settling time for statistically detecting healthy antenna behaviour may be relatively long, which may be critical in constantly changing network topologies with frequent communications links disruption.

Platooning applications constitute a particular use case scenario of vehicular communications. Ploeg et al. [23] address the problem of faulty links in a platoon of vehicles. A safe distance between the members of the platoon is continuously computed by taking into consideration the availability of sensor data and the communications link performance. This safe distance is employed by the cooperative adaptive cruise control (CACC) system according to a graceful degradation scheme that adjusts the settings of CACC to keep as much functionality as possible, even in the presence of faults, but always guaranteeing string stability in the platoon. Moreover, two different network topologies can be applied, depending on the time delay of the communications link. If this delay exceeds a predefined threshold value, the platoon service switches from a one-vehicle look-ahead topology to a two-vehicle look-ahead configuration. This fault-tolerance strategy can only be applied if the delay time is not excessively large, otherwise, wireless communications should not be employed in order to preserve string stability.

The described fault-tolerant scheme targets only a particular application, i.e., vehicle platooning, being tied to a concrete network topology and thus not very useful to other use case scenarios. The fault model only takes into account large delay values that may affect the timeliness of the communications links, not considering faults in the value domain of the transmitted packets. Additionally, only numerical results are provided, which makes it difficult to evaluate the performance of the system under real environments with adverse conditions.

In [24], Fathollahnejad et al. propose a synchronous group formation (GF) algorithm to enhance self-organizing vehicular applications under the presence of an unbounded number of asymmetric communication failures. The main goal of the GF algorithm is to achieve agreement, or at least to reduce the probability of unsafe disagreement, on the membership of a cooperative ITS application, e.g., virtual traffic light (VTL) systems. A decision mechanism is employed by each member node (vehicle) to identify the other nodes in the group at each moment in time. The mechanism relies on the utilization of an extra component, designated as *oracle*. These *oracles* are local devices present in each node and are responsible for detecting the remaining participants in the group. The obtained simulation results show that when the local *oracles* provide a correct estimate of the group formation, only safe disagreement scenarios may occur. However, when the *oracles* underestimate the total number of nodes, unsafe disagreement situations may happen and the likelihood of such scenarios increases with the probability of receive omissions in the communications channel.

This work excludes process failures, only dealing with faults in the communications links and more specifically just receive omissions faults, so faults in the value domain are also outside of the scope of the fault model. Moreover, communications overhead for leader election, leader handover or VTL group formation protocol is not discussed and analysed, and the leader election and leader handover mechanism are not yet designed, being part of future work. Finally, the evaluation section only presents numerical results, there are no experiments conducted in more realistic traffic/network simulation or test case environments.

Bhoi and Khilar [25] introduce a fault-tolerant routing protocol for vehicular ad-hoc communications in urban environments. A fault detection technique is used by the vehicle itself to detect if its own operation is fault free or not. If a faulty behaviour is identified, the on-board unit (OBU) does not participate any longer in the routing process. The fault detection mechanism targets soft faults, i.e., erroneous behaviour in the OBU devices causing the generation of incorrect data for a long period of time. This may be caused by high noise affecting the node's operation, making it still able to compute, send and receive information. However, beaconing data transmitted by the vehicle cannot be considered valid, being that this information (position, speed, etc.) is indispensable for hop selection in the routing algorithm. For that reason, these nodes are automatically excluded from the routing process by self-detecting these soft faults, through the analysis of the RSSI values from the received messages and the location coordinates provided by the neighbouring nodes. The proposed protocol provides good results in terms of end-to-end delay, path length and false alarm rate.

The proposed routing protocol just takes into account faults in the value domain, e.g., incorrect data in the position or speed information, not considering the possibility of nodes introducing timing faults, such as large delay values. It is also assumed that the faulty vehicles always provide incorrect data and only by accident the information may be correct. This simplifies the fault detection mechanism but could be a not very realistic situation, since nodes may present intermittent faults that only arise in some occasions. The overhead of exchanging decision messages regarding the state (soft faulty or fault free) of neighbouring vehicles is neither discussed nor analysed. Furthermore, despite the decentralized fault detection scheme, the routing and path value calculation algorithm depends on RSU nodes, which are single points of failure in this forwarding scheme.

The work of Mourad Elhadef [26] suggests the utilization of a primary-backup approach for the design of a fault-tolerant intersection control algorithm. The VTL system is based on a centralized solution, with an RSU controller responsible for coordinating all traffic crossing the intersection. The controller manages the vehicles approaching the site, by granting or denying access to the intersection, in order to guarantee safety, liveliness and fairness, while at the same time maximizing traffic throughput. Both the primary and the backup controllers are constantly synchronizing with each other, so that the backup unit can always be kept updated with all the necessary traffic information. Only permanent crash failures are considered in the fault model. Whenever the main controller stops sending and receiving messages (a keep alive timer is used to detect if the primary node is down), the backup unit takes control of the intersection.

This fault-tolerant intersection control algorithm has the drawback of not dealing with intermittent faults and message errors in the value domain, by only considering permanent crash failures. Besides that, it focus on a specific application (intersection management), while a more generic solution could be devised with the same primary-backup approach for master nodes of vehicular networks that rely on centralized or hybrid architectures. Furthermore, no evaluation of the proposed scheme is carried out, namely in terms of recovery delay after primary replica failure.

An RSU-backup replication scheme is proposed by Almeida et al. [27] in the scope of a fault-tolerant infrastructure-based architecture for vehicular networks. In this framework, the RSUs behave as the masters of the network, controlling time slot scheduling of the OBUs and admission control policies. They have a crucial role in the network operation, acting as single point of failure, and therefore, any fault affecting these nodes may cause a disruption in the time-sensitive communications for safety-critical applications. The work introduces a full replication scheme, where a backup node executes the exact same processes as the primary fail-silent replica. This parallel operation allows the system to perform a very fast recovery procedure in case of failure of the primary node. As a result, the real-time communications protocol does not suffer any discontinuity even in the presence of network faults, thus enhancing the overall dependability of vehicular system.

The replication mechanism proposed in this work consists in a cross-layer approach, since it requires the duplication of the entire RSU node from the physical up to the application layer. Unfortunately, there is no cost-benefit analysis for this solution, since the complete replication of hardware and software parts is expensive and could be compared against other possibilities, such as the option for non fail-silent RSUs and the use of backup ones operating in another channel frequency. Moreover, the deterministic MAC protocol that is on the basis of this architecture also needs some changes to the standard vehicular communications protocols, so the solution cannot be seamlessly implemented on COTS components. Finally, the fault-tolerant system was only tested in a controlled laboratory environment. Some testing in close-to-real world scenarios could give an improved understanding of system's reliability under the presence of hardware, software or communications channel faults.

In [28], Medani et al. develop a time synchronization strategy for the nodes of vehicular networks. Clock synchronization is essential to support the correct operation of safety-critical applications in road traffic environments (e.g., for event causality, medium access control or security purposes). The proposed method, named Offset Table Robust Broadcasting, attains high accuracy and presents fault-tolerant capabilities so that every node is aware of neighbouring clock times and is able to synchronize its communications with other nodes. The clock offsets among several nodes are computed using a round-trip time mechanism and acknowledgement messages are exchanged to ensure that the offset table delivery reaches all nodes.

In order to achieve higher clock synchronization, the proposed scheme introduces some communications overhead, both for the transporter node selection and cluster formation but also for the entire synchronization process that involves collecting timing information, calculate offsets table and disseminate the computed data. Additionally, it requires the modification of the clock synchronization source in the node, which may be sometimes difficult to implement in COTS components. Finally, it would be interesting to perform some field trials evaluation, in order to have real GPS errors and uncertainties in the synchronization process.

A multipath routing protocol is proposed by Devangavi et al. [29], in order to enhance reliability and fault tolerance. Multiple paths are computed from source to the destination node based on Bezier curves. These curves are traced by the parent RSU according to the geographical location of the different nodes in the network on a multi-hop coverage area. The calculation of these paths is also based on several parameters, such as the available bandwidth in the network, the data size to be transmitted and the distances from source to destination. The distinct paths are then prioritized and utilized to forward the information to the destination vehicle, introducing a flexible degree of redundancy in message transmission. The proposed solution was evaluated taking as example the city of Bangalore and the simulation results obtained in NS-2 proved the superior performance of the protocol in comparison with other solutions in the literature and with respect to transmission time and packet delivery ratio.

The proposed protocol is based on a centralized architecture, where RSU nodes are responsible for multipath finding process and network management tasks. However, these nodes are single point failures that in case of permanent crash, disrupt network operation. Furthermore, it is assumed that every vehicle is always connected to at least one RSU, which limits the applicability of the proposed solution in real-world scenarios. It should also be noted that a prioritized path list must be computed for every source-destination pair of vehicles involved in message exchange, which introduces a significant amount of communications overhead that is not evaluated in the simulation experiments.

In [30], Younes et al. propose the FT-PR protocol, a fault-tolerant path recommendation system. In this work, vehicles within a reporting area are responsible for disseminating the traffic characteristics of a road segment. The process is cumulative, since a road segment can have multiple reporting areas. Transmitting vehicles gather information on surrounding clusters and a report is completed as soon as it encompasses the entire road segment. Road-side units (RSUs), assumed to be located at each road intersection, exchange information with each other and calculate the best road segments for specific destinations. RSUs start broadcasting destinations and the best turn towards them. From this part forward, the process is iterative, vehicles entering a road segment receive the path recommendation information and may progress towards the road network. Different techniques are used to improve the robustness of the system, For instance, in order to enhance the traffic collection phase, vehicles can request updates by disseminating vehicles missing in the report description. Furthermore, in case of an RSU error, the nearby RSUs can assume their roles and transmit their information. Additionally a vehicle can retransmit messages from an RSU, in order to increase the RSU communication radius.

The main contribution of this work consists in the designation of redundant routing paths for multi-hop communications, in order to compensate for RSU failures. The solution does not address the issue of a vehicle reaching a specific RSU node, e.g., one responsible for a safety-critical task, such as controlling a road intersection. If the destination RSU is faulty, no redundant node is available to perform the expected task. Another drawback of the proposed protocol is that the selection of a cluster head for each reported area, a constantly dynamic process, may introduce a significant amount of overhead, which may be critical in terms of delay. Nothing is mentioned in the article about how this selection process is conducted. In terms of results, by using FT-PR protocol, vehicles were able to obtain the optimal path even if 40% of installed RSUs failed to process or forward the advertisement messages. However, the obtained simulation metrics are not generic and were only evaluated for specific layout scenarios. More realistic situations should be considered for further analysis.

4.3. Information Redundancy

Finally, with respect to information redundancy mechanisms, the identified solutions are based on network coding techniques. For instance, the work of Kumar and Dave [8] introduces a decentralized method that provides reliable and scalable vehicular communications, independently of the traffic density level. The solution employs network coding and random walks in order to deal with the constantly changing topologies, varying vehicle density and unreliable channel conditions of vehicular networks. Raptor codes, which are characterized by its low complexity and thus fast decoding, are used to encode and disseminate information in a completely distributed manner, providing better fault tolerance. In this scheme, a vehicle transmits its data to a random set of neighbouring vehicles and then each vehicle only encodes the information it has received. Posteriorly, a receiving vehicle can efficiently decode the transmitted data by collecting a sufficient amount of data blocks from the network nodes it interacts with while moving. Random walks are utilized to disseminate the data, avoiding the need for supporting a generic layer of routing protocols. The performance evaluation of the proposed scheme is evaluated through simulation and the results are compared against the simple broadcast framework (store and forward strategy) and other related work in the literature. Data overhead and average end-to-end delay are kept low for different traffic densities and data sending rates, while network reachability and packet delivery ratio are improved in comparison with the other analysed solutions.

The suggested scheme requires changes to the standard physical and routing layers, due to encoding/decoding process and data forwarding mechanisms. As a result, it cannot be directly deployed on top of COTS equipment, by just operating at the application level. The method also introduces some communications overhead in exchange for increased redundancy, but completely manageable according to the simulation results. Finally, the performance of the proposed method could be further evaluated by using more realistic traffic models in the simulation tools or by taking results in field trial scenarios.

Eze et al. [11] also propose an innovative communications scheme based on the network coding concept, named Coding Aided Retransmission-Based Error Recovery (CARER). The goal is to improve broadcast reliability and timely delivery of messages with lower number of retransmissions. In this scheme, each node performs an exclusive OR (XOR) operation on a set of both received and generated packets and then send these encoded messages to all the vehicles in a one-hop distance. The advantage of this technique over simply broadcasting the raw packets is that it allows nodes to recover lost frames with low communications overhead. Additionally, the protocol uses a location-aware algorithm that selects an appropriate vehicle for rebroadcasting the encoded packet towards the desired propagation direction. The traditional Request-to-Broadcast/Clear-to-Broadcast (RTB/CTB) handshake is used to overcome hidden node problem and reduce collisions in the wireless medium. An analytical model was developed and simulation tests were performed to evaluate the performance of the protocol. The results show an increased packet recovery probability and a lower packet collision probability when compared to the simple repetition based error recovery scheme with no network coding mechanism involved.

The protocol only takes into consideration faults in the communications links, not dealing with node failures, which may be critical for the operation of the location-aware algorithm that selects a specific node to retransmit the encoded packets. Moreover, the utilization of the RTB/CTB handsahke introduces some communications overhead that is analysed but not evaluated in the experimental results section, in terms of total end-to-end delay. Finally, the protocol presents satisfactory results for a straight road scenario with no intersection, which is typically not the case in urban environments, where an unappropriate node can be easily selected to retransmit the encoded packets, i.e., forwarding them in the wrong direction. These more complex and dense topologies must be further evaluated.

Ali et al. [14] addresses one of the main challenges in vehicular communications, the degradation of radio propagation. It proposes a protocol for code-relaying information at road intersections. In the proposed scheme, a station can mediate messages between two other vehicles, B and C, by broadcasting a XOR of the received periodic status messages from both B and C. B and C can retrieve each other's messages by performing the same operation and removing their own messages. The protocol was implemented in NS3 and tested using SUMO, showing that the impact of path loss, fading and shadowing can be highly mitigated by the before-mentioned technique. In particular, it showed that using coded relay, a station could expect higher packet delivery rates and smaller latency (per-instance). Unfortunately, due to the relay mechanism, it takes more time to perform successful reception.

The proposed solution handles communications links faults, such as packet drops, but not node failures or faults in the value domain, e.g., incorrect data messages. This is particularly important for the proposed relaying at road intersection, where a relay node assists the message exchange. These relays are single point of failures in the communications framework, that in case of faulty behaviour, will compromise the packet forwarding operation. Furthermore, this is even more challenging when the relay node is not an RSU but an opportunistic vehicle that, due to its speed, may only create temporary and unstable connection.

4.4. Discussion

Table 2 provides a comparison among different aspects of the selected documents. A description of the proposed fault-tolerance technique is depicted, including the main applicability, the method used, the target scenario (urban, highway or both) and the structure or architecture of the proposal (centralized, distributed or hybrid). With respect to the fault-tolerance or fault-detection method itself, the technique is classified according to its type, the protocol layer in which operates and the methodology to provide error handling, fault handling or both [2]. Finally, the publications are compared in terms of the evaluation performed, namely the parameters measured, and type of analysis carried out (analytical model, numerical, simulation or real testbed).

Most of the articles provide decentralized solutions, but there are also some hybrid and centralized architectures. Regarding the classification of the protocol stack layer in which the fault-tolerant behaviour is provided, this is not a very straightforward task, since many solutions have an impact at multiple levels. When it is clear that a contribution has a cross-layer approach, the different layers are mentioned, otherwise the most suitable protocol layer is selected for trying to categorize the proposed solution. The protocol layers division is based on the standard protocol stack for wireless vehicular communications [32], namely splitting into the physical, MAC and LLC, network, transport and application layers.

The fault tolerance methods were also analysed in terms of the error handling and fault handling mechanisms used. With respect to the error handling techniques, most of the articles proposed a compensation scheme, in which the erroneous state contains enough redundancy to enable error to be masked, however, in some cases, a rollforward strategy was followed, where the system jumps to a new state without detected errors. For the fault handling part, reconfiguration of the system was typically employed, either switching in spare components or reassigning tasks among non-failed components, while in some solutions isolation and reinitialization mechanisms were also included. Regarding the evaluation process of the proposed solutions, a set of different parameters were evaluated in each case, but most of them relied on analytical models, numerical analysis or simulation results. Only in two articles, a real testbed was used.

In summary, most of the existing fault-tolerance techniques proposed in the literature only take into consideration specific vehicular communications applications, not being designed to support the full range of services enabled by these networks. Moreover, these solutions typically cover just a small number of faults affecting the vehicular communications system, focusing for instance on a particular layer of the protocol stack, e.g., routing protocol. As a result, none of the solutions can fully address the stringent dependability requirements of such safety-critical wireless systems, namely the high reliability and availability levels. For that to happen, an integrated approach of the entire network architecture needs to be taken into account, with an extended fault coverage of the system's operation. In addition to this, more practical implementations of fault-tolerant methods need to be developed, in order to evaluate the proposed mechanisms with field experimental results.

Table 2. Comparison of selected documents.

	Description						Fault Detection/Fault Tolerance				Evaluation			
	Applicability	Method	Scenario	Structure	Type	Layer	Error Handling	Fault Handling	Parameters	Analytical Model	Numerical Analysis	Simulation	Real Test-Case
Jonsson et al. [5]	Platooning	Message Retransmission	Generic	Centralized	Temporal	Transport	Compensation	—	• Message Error Rate • Channel Busy Time	✓	✗	Matlab	✗
Böhm and Kunert [6]	Platooning	Message Retransmission	Generic	Centralized	Temporal	Transport	Compensation	—	• Packet Delivery Ratio • Data Age	✗	✗	MatLab	✗
Savic et al. [12]	Intersection Control	Message Retransmission	Urban	Decentralized	Temporal	Transport	Compensation	—	• Packet Delivery Ratio • Average Delay	✓	✓	✗	✗
Sawade et al. [15]	Coordinated Maneuvers	Bidirectional Stateful Communication + Distributed Consensus	Highway	Decentralized	Temporal	Transport	Compensation	—	• Unstable Session Ratio	✗	✗	VSimRTI	✗
Nguyen et al. [17]	Distributed ITS Applications	Message Retransmission	Highway	Centralized	Temporal	Transport	Compensation	—	• Packet Delivery Ratio	✓	✓	MatLab	✗
Matthiesen et al. [7]	Service Replication	Petri Networks + Markov Chains	Generic	Decentralized	Spatial	Network	Compensation	Reconfiguration	• Service Availability • Group Consistency	✓	✓	✗	✗
Cambruzzi et al. [14]	Distributed ITS Applications	Failure Detection System	Generic	Decentralized	Spatial	Application	—	—	• Percentage of False Suspicions • Average Time of Detection • Average Time of Recovery	✗	✗	OMNET++	✗
Abroungui et al. [13]	Service Discovery	Spanning Tree Reconstruction	Generic	Centralized	Spatial	Application	Rollforward	Reconfiguration	• Recovery Success Rate • Used Bandwidth • Response Time	✓	✓	✗	✗
Chang and Wang [16]	Message Broadcast	Relay Candidates Selection	Highway	Hybrid	Spatial	Network	Compensation	Reconfiguration	• Number of Messages • Transmission Delay • Penetration Rate	✗	✗	NS-2	✗
Le Lann [18]	Platooning	Cohorts + Proxy Sets + Zebra protocols suite	Generic	Hybrid	Spatial	Network	Compensation	Reconfiguration Reinitialization	• Worst Case Termination Time	✓	✓	✗	✗
Sanderson and Pitt [19]	Distributed Databases	Institutionalised Consensus	Generic	Hybrid	Spatial	Application	Compensation	Reconfiguration Reinitialization	• Packet Delivery Ratio • Data Overhead • Average Delay Network • Reachability	✓	✗	✗	✗
Aljeri et al. [20]	Roadside ITS Applications	Spanning Tree Reconstruction	Generic	Centralized	Spatial	Application	Rollforward	Isolation Reinitialization	• Number of Packets • Diagnosis Latency	✗	✗	NS-2	✗
Casimiro et al. [21]	Vehicle Coordination	Failure Detection System + Graceful Degradation	Generic	Decentralized	Spatial	MAC and Network	Rollforward	Reconfiguration	—	✗	✗	✗	✗
Worrall et al. [22]	Distributed ITS Applications	Radio Redundancy + Machine Learning	Generic	Decentralized	Spatial	Physical	Compensation	Reconfiguration	• Antenna Performance	✗	✗	✗	✓
Ploeg et al. [23]	Platooning	Communication Topology Adaptation	Generic	Decentralized	Spatial	Network	Compensation	Reconfiguration	—	✓	✓	✗	✗

Table 2. Cont.

		Description				Fault Detection/Fault Tolerance			Evaluation		
Fatollahnejad et al. [25]	Group Formation	Consensus	Generic	Centralized	Spatial	Application	Rollforward	—	• Probability of Safe Disagreement • Probability of Unsafe Disagreement	✓	✓	✗
Bhoi and Khilar [26]	Multi-hop Routing	Self Soft Fault Detection	Urban	Decentralized	Spatial	Network	Compensation	Isolation Reconfiguration	• Fault Detection Rate • False Alarm Rate • End-to-end Delay • Number of gaps • Number of hops	✓	✗	MatLab	✗
Elhadef [26]	Intersection Control	Controller Redundancy	Urban	Centralized	Spatial	Application	Compensation	Reconfiguration	—	✗	✗	✗	✗
Almeida et al. [27]	Distributed ITS Applications	RSU Replication Scheme	Generic	Centralized	Spatial	Application	Compensation	Isolation	• Backup Replica Offset	✗	✗	✗	✓
Medani et al. [28]	Time Synchronization	Offsets Table Broadcasting Protocol	Generic	Hybrid	Spatial	Network	Rollforward	Reconfiguration Reinitialization	• Message Complexity • Convergence Time • Synchronization Rate	✓	✓	NS-2 VanetMobiSim	✗
Devangavi et al. [29]	Multi-hop Routing	Bezier Curves + Path Redundancy	Generic	Centralized	Spatial	Network	Compensation	—	• Transmission Time • Packet Delivery Ratio	✗	✗	NS-2	✗
Younes et al. [30]	Road Path Recommendation	Path Redundancy	Urban	Decentralized	Spatial	Network	Compensation	—	• Average Traveling Time • Average Traveling Distance • Number of Sent Packets • Average Delay Time • Un-tolerated Scenarios • Correctness	✗	✗	NS-2 + SUMO	✗
Kumar and Dave [9]	Message Broadcast	Network Coding + Raptor Codes + Markov Chains	Generic	Decentralized	Information	Network	Compensation	—	• Packet Delivery Ratio • Data Overhead • Average Delay Network Reachability	✗	✗	NS-2 SUMO MOVE	✗
Eze et al. [11]	Message Broadcast	Network Coding + Message Retransmission	Highway	Decentralized	Information	Network and Transport	Compensation	—	• Packet Recovery Probability • Packet Collision Probability	✓	✗	NS-2 Bonn-Motion tool	✗
Ali et al. [4]	Intersection Control	Network Coding + Message Relaying	Urban	Decentralized	Information	Network	Compensation	—	• Expected per-instance latency • Time to successful reception • Packet Delivery Rate	✓	✗	NS-3 SUMO	✗

112

5. Conclusions

In this work, a systematic and comprehensive survey on fault tolerance techniques for wireless vehicular networks was conducted. In summary, there are not many research works in the area of fault-tolerance specifically focusing on wireless vehicular communications. However, an increasing trend can be observed in the recent years, with more protocols, mechanisms and architectures being proposed in order to enhance the dependability attributes of wireless vehicular networks. A systematic process to select publications from a large dataset was followed, choosing the ones that are more relevant and specifically focused on fault-tolerance in wireless vehicular communications. The analysed papers show that the development of safety-critical applications in such dynamic environments require a careful planning that preserves the system's flexibility and real-time guarantees while providing fault-tolerance capabilities. As a conclusion, there is still a shortage of strategies to completely fulfil the operation of dependable vehicular networks. Nevertheless, this is a crucial requirement of vehicular communications, namely for the safety-critical applications supported by these networks.

Author Contributions: Conceptualization and methodology, J.A. and J.F.; formal analysis, investigation, writing–original draft preparation and visualization, J.A. and J.R.; writing–review and editing, M.A. and J.F.; supervision and project administration and funding acquisition, J.F.

Funding: This work is supported by the European Regional Development Fund (FEDER), through the Competitiveness and Internationalization Operational Programme (COMPETE 2020) of the Portugal 2020 framework [Project TRUST with Nr. 037930 (POCI-01-0247-FEDER-037930)].

Conflicts of Interest: The authors declare no conflict of interest. The funders had no role in the design of the study; in the collection, analyses, or interpretation of data; in the writing of the manuscript, or in the decision to publish the results.

References

1. Gärtner, F.C. Fundamentals of Fault-tolerant Distributed Computing in Asynchronous Environments. *ACM Comput. Surv.* **1999**, *31*, 1–26. [CrossRef]
2. Avizienis, A.; Laprie, J.C.; Randell, B.; Landwehr, C. Basic concepts and taxonomy of dependable and secure computing. *IEEE Trans. Dependable Secur. Comput.* **2004**, *1*, 11–33. [CrossRef]
3. Lima, A.; Rocha, F.; Völp, M.; Esteves-Veríssimo, P. Towards Safe and Secure Autonomous and Cooperative Vehicle Ecosystems. In Proceedings of the 2nd ACM Workshop on Cyber-Physical Systems Security and Privacy, CPS-SPC '16, Vienna, Austria, 28 October 2016; ACM: New York, NY, USA, 2016; pp. 59–70. [CrossRef]
4. Jhumka, A.; Kulkarni, S. On the Design of Mobility-Tolerant TDMA-Based Media Access Control (MAC) Protocol for Mobile Sensor Networks. In *Distributed Computing and Internet Technology*; Janowski, T., Mohanty, H., Eds.; Springer: Berlin/Heidelberg, Germany, 2007; pp. 42–53.
5. Moher, D.; Liberati, A.; Tetzlaff, J.; Altman, D.G. Preferred reporting items for systematic reviews and meta-analyses: The PRISMA statement. *BMJ* **2009**, *339*. [CrossRef] [PubMed]
6. Jonsson, M.; Kunert, K.; Böhm, A. *Increased Communication Reliability for Delay-Sensitive Platooning Applications on Top of IEEE 802.11p*, Proceedings of the Communication Technologies for Vehicles: 5th International Workshop, Nets4Cars/Nets4Trains 2013, Villeneuve d'Ascq, France, 14–15 May 2013; Berbineau, M., Jonsson, M., Bonnin, J.M., Cherkaoui, S., Aguado, M., Rico-Garcia, C., Ghannoum, H., Mehmood, R., Vinel, A., Eds.; Springer: Berlin/Heidelberg, Germany, 2013; pp. 121–135. [CrossRef]
7. Matthiesen, E.V.; Hamouda, O.; Kaâniche, M.; Schwefel, H.P. Dependability Evaluation of a Replication Service for Mobile Applications in Dynamic Ad-Hoc Networks. In *Service Availability*; Nanya, T., Maruyama, F., Patariczca, A., Malek, M., Eds.; Springer: Berlin/Heidelberg, Germany, 2008; pp. 171–186.
8. Kumar, R.; Dave, M. DDDRC: Decentralised data dissemination in VANET using raptor codes. *Int. J. Electron.* **2015**, *102*, 946–966. [CrossRef]
9. Böhm, A.; Kunert, K. Data age based MAC scheme for fast and reliable communication within and between platoons of vehicles. In Proceedings of the 2016 IEEE 12th International Conference on Wireless and Mobile Computing, Networking and Communications (WiMob), New York, NY, USA, 17–19 October 2016; pp. 1–9. [CrossRef]

10. Cambruzzi, E.; Farines, J.M.; Macedo, R.J.; Kraus, W. An adaptive failure detection system for Vehicular Ad-hoc Networks. In Proceedings of the 2010 IEEE Intelligent Vehicles Symposium, San Diego, CA, USA, 21–24 June 2010; pp. 603–608. [CrossRef]
11. Eze, E.C.; Zhang, S.; Liu, E. Improving Reliability of Message Broadcast over Internet of Vehicles (IoVs). In Proceedings of the 2015 IEEE International Conference on Computer and Information Technology; Ubiquitous Computing and Communications; Dependable, Autonomic and Secure Computing; Pervasive Intelligence and Computing, Liverpool, UK, 26–28 October 2015; pp. 2321–2328. [CrossRef]
12. Savic, V.; Schiller, E.M.; Papatriantafilou, M. Distributed algorithm for collision avoidance at road intersections in the presence of communication failures. In Proceedings of the 2017 IEEE Intelligent Vehicles Symposium (IV), Los Angeles, CA, USA, 11–14 June 2017; pp. 1005–1012. [CrossRef]
13. Abrougui, K.; Boukerche, A.; Ramadan, H. Performance evaluation of an efficient fault tolerant service discovery protocol for vehicular networks. *J. Netw. Comput. Appl.* **2012**, *35*, 1424–1435. [CrossRef]
14. Ali, G.G.M.N.; Noor-A-Rahim, M.; Chong, P.H.J.; Guan, Y.L. Analysis and Improvement of Reliability Through Coding for Safety Message Broadcasting in Urban Vehicular Networks. *IEEE Trans. Veh. Technol.* **2018**, *67*, 6774–6787. [CrossRef]
15. Sawade, O.; Schulze, M.; Radusch, I. Robust Communication for Cooperative Driving Maneuvers. *IEEE Intell. Transp. Syst. Mag.* **2018**, *10*, 159–169. [CrossRef]
16. Chang, Y.C.; Wang, T.P. A fault-tolerant broadcast protocol for reliable alert message delivery in vehicular wireless networks. In Proceedings of the 7th International Conference on Communications and Networking in China, Kun Ming, China, 8–10 August 2012; pp. 475–480. [CrossRef]
17. Nguyen, V.; Khanh, T.T.; Oo, T.Z.; Tran, N.H.; Huh, E.; Hong, C.S. A Cooperative and Reliable RSU-Assisted IEEE 802.11P-Based Multi-Channel MAC Protocol for VANETs. *IEEE Access* **2019**, *7*, 107576–107590. [CrossRef]
18. Lann, G.L. On the Power of Cohorts—Multipoint Protocols for Fast and Reliable Safety-Critical Communications in Intelligent Vehicular Networks. In Proceedings of the 2012 International Conference on Connected Vehicles and Expo (ICCVE), Beijing, China, 12–16 December 2012; pp. 35–42. [CrossRef]
19. Sanderson, D.; Pitt, J. Institutionalised Consensus in Vehicular Networks: Executable Specification and Empirical Validation. In Proceedings of the 2012 IEEE Sixth International Conference on Self-Adaptive and Self-Organizing Systems Workshops, Lyon, France, 10–14 September 2012; pp. 71–76. [CrossRef]
20. Aljeri, N.; Almulla, M.; Boukerche, A. An Efficient Fault Detection and Diagnosis Protocolfor Vehicular Networks. In Proceedings of the Third ACM International Symposium on Design and Analysis of Intelligent Vehicular Networks and Applications, Barcelona, Spain, 3–8 November 2013; ACM: New York, NY, USA, 2013; DIVANet '13, pp. 23–30. [CrossRef]
21. Casimiro, A.; Kaiser, J.; Schiller, E.M.; Costa, P.; Parizi, J.; Johansson, R.; Librino, R. The KARYON project: Predictable and safe coordination in cooperative vehicular systems. In Proceedings of the 2013 43rd Annual IEEE/IFIP Conference on Dependable Systems and Networks Workshop (DSN-W), Budapest, Hungary, 24–27 June 2013; pp. 1–12. [CrossRef]
22. Worrall, S.; Agamennoni, G.; Ward, J.; Nebot, E. Fault Detection for Vehicular Ad Hoc Wireless Networks. *IEEE Intell. Transp. Syst. Mag.* **2014**, *6*, 34–44. [CrossRef]
23. Ploeg, J.; van de Wouw, N.; Nijmeijer, H. Fault Tolerance of Cooperative Vehicle Platoons Subject to Communication Delay. *IFAC-PapersOnLine* **2015**, *48*, 352–357. [CrossRef]
24. Fathollahnejad, N.; Pathan, R.; Karlsson, J. On the Probability of Unsafe Disagreement in Group Formation Algorithms for Vehicular Ad Hoc Networks. In Proceedings of the 2015 11th European Dependable Computing Conference (EDCC), Paris, France, 7–11 September 2015; pp. 256–267. [CrossRef]
25. Bhoi, S.; Khilar, P. Self soft fault detection based routing protocol for vehicular ad hoc network in city environment. *Wirel. Netw.* **2016**, *22*, 285–305. [CrossRef]
26. Elhadef, M. A Fault-Tolerant Intersection Control Algorithm Under the Connected Intelligent Vehicles Environment. In *Advanced Multimedia and Ubiquitous Engineering*; Park, J., Jin, H., Jeong, Y.S., Khan, M., Eds.; Lecture Notes in Electrical Engineering; Springer: Singapore, 2016; Volume 393, pp. 243–253. [CrossRef]
27. Almeida, J.; Ferreira, J.; Oliveira, A.S.R. An RSU Replication Scheme for Dependable Wireless Vehicular Networks. In Proceedings of the 2016 12th European Dependable Computing Conference (EDCC), Gothenburg, Sweden, 5–9 September 2016; pp. 229–240. [CrossRef]

28. Medani, K.; Aliouat, M.; Aliouat, Z. Fault tolerant time synchronization using offsets table robust broadcasting protocol for vehicular ad hoc networks. *AEU Int. J. Electron. Commun.* **2017**, *81*, 192–204. [CrossRef]
29. Devangavi, A.D.; Gupta, R. Bezier Curve based Multipath Routing in VANET. In Proceedings of the 2018 Second International Conference on Advances in Electronics, Computers and Communications (ICAECC), Bangalore, India, 9–10 February 2018; pp. 1–5. [CrossRef]
30. Younes, M.B.; Boukerche, A. A performance evaluation of a fault-tolerant path recommendation protocol for smart transportation system. *Wirel. Netw.* **2018**, *24*, 345–360. [CrossRef]
31. Lamport, L. The Part-time Parliament. *ACM Trans. Comput. Syst.* **1998**, *16*, 133–169. [CrossRef]
32. Kenney, J. Dedicated Short-Range Communications (DSRC) Standards in the United States. *IEEE Proc.* **2011**, *99*, 1162–1182. [CrossRef]

 © 2019 by the authors. Licensee MDPI, Basel, Switzerland. This article is an open access article distributed under the terms and conditions of the Creative Commons Attribution (CC BY) license (http://creativecommons.org/licenses/by/4.0/).

MDPI
St. Alban-Anlage 66
4052 Basel
Switzerland
Tel. +41 61 683 77 34
Fax +41 61 302 89 18
www.mdpi.com

Electronics Editorial Office
E-mail: electronics@mdpi.com
www.mdpi.com/journal/electronics

www.ingramcontent.com/pod-product-compliance
Lightning Source LLC
LaVergne TN
LVHW071443100526
838202LV00088B/6780